Enrique Kremers

Modelling and Simulation of Electrical Energy Systems through a Complex Systems Approach using Agent-Based Models

Modelling and Simulation of Electrical Energy Systems through a Complex Systems Approach using Agent-Based Models

by
Enrique Kremers

KIT Scientific Publishing

This thesis was supervised by Dr. Oscar Barambones Caramazana, Automatic Control and System Engineering Department at Universidad del País Vasco (UPV/EHU) in Spain as an International Doctorate (Doctor Internacional). This modality prescribes a research stay abroad and the composition of the dissertation or a part of it in a foreign language. The international doctorate was elaborated in collaboration with the European Institute for Energy Research (EIFER) in Karlsruhe, Germany, which is a joint institute founded by both Karlsruhe Institute of Technology (KIT) and Electricité de France (EDF).

Acta de Grado de Doctor
Acta de Defensa de Tesis Doctoral

El Tribunal designado por la Subcomisión de Doctorado de la UPV/EHU para calificar la Tesis Doctoral arriba indicada y reunido en el dia de la fecha, una vez efectuada la defensa por el doctorando y contestadas las objeciones y/o sugerencias que se le han formulado, ha otorgado por unanimidad la calificación de: APTO CUM LAUDE.

En Vitoria-Gasteiz, a 27 de Julio de 2012.

El Presidente: Dr. Manuel de la Sen
El Secretario: Dr. José María González de Durana
Vocal 1°: Dr. Francisco Maciá Pérez
Vocal 2°: Dr. José Manuel Andújar Márquez
Vocal 3°: Dr. Kevin McKoen

Dissertation, Universidad del Pais Vasco
Escuela Universitaria de Ingeniería de Vitoria
Departamento Ingeniería de Sistemas y Automática, 2012

Impressum

Karlsruher Institut für Technologie (KIT), KIT Scientific Publishing
Straße am Forum 2, D-76131 Karlsruhe
www.ksp.kit.edu

KIT – Universität des Landes Baden-Württemberg und
nationales Forschungszentrum in der Helmholtz-Gemeinschaft

KIT Scientific Publishing 2013
Print on Demand

ISBN 978-3-86644-946-6

Abstract

Our world is facing a significant challenge from climate change and global warming, coupled with an increased awareness about the importance of preserving the environment. This challenge calls for optimising our resources and developing in a more sustainable way. Energy, as one of the main contributors to air emissions and pollution, holds great potential for improving this area. One of the results of moving towards sustainability is the general trend of introducing renewable energy sources in industrialised countries. This implies a great change in the structure of energy systems, moving away from a centralised and hierarchical energy system towards a new system influenced by diverse actors. Along with political decisions, such as the deregulation of the energy sector a paradigm shift in the energy system has been initiated.

The electrical energy system is traditionally an interconnected, large scale system with dynamic behaviour over time; it is composed of networks at different levels spread over vast geographical areas. Different participants from these areas, each with their own range of local interests and objectives, all interact with the energy system. These factors indicate that the energy system can be approached as a *complex system*. Complexity science is an emerging interdisciplinary field of research that has been mainly studied in the social sciences, biology and physics. However, complexity, as such, is not restricted to these areas as a subject of research. It aims to better understand and analyse the processes of both natural and man-made systems which are composed of many interacting entities at different scales. Complexity can be found in the collective behaviour of large number of entities at a system level, swarms of birds or ant trails for example. This behaviour at the collective level, cannot, however, be directly inferred from the behaviour of the individual parts of the system. Complexity science is also closely related to network theory, which is used to describe relations or interactions among the entities. So, for example, the topology of an electrical system is found to be scale-free, which means that there are many nodes with few connections, but only some very well connected ones. Another application of complexity related to electrical energy systems is the study of resilience of the network against targeted attacks, based on its topology. However, a *complex* approach in the energy domain is still marginal. In this thesis, the relevance and interest of a complex systems approach is discussed, as there seem to be many parallels to other systems already studied.

The hypothesis of how such an approach could help to better understand the behaviour of energy systems is initially treated from a theoretical point of view. In a second stage, the application of the approach is illustrated through some examples of modelling and simulation.

One of the ways of studying complex systems is through modelling and simulation, which are used as tools to represent these systems in a virtual environment. Current advances in computing performance (which has been a major constraint in this field for some time) allow for the simulation these kinds of systems within reasonable time horizons.

One of the tools for simulating complex systems is agent-based models. This individual-centric approach is based on autonomous entities that can interact with each other, thus modelling the system in a disaggregated way. Agent-based models can be coupled with other modelling methods, such as continuous models and discrete events, which can be embedded or run in parallel to the multi-agent system. When representing the electrical energy system in a systemic and multi-layered way, it is treated as a true socio-technical system, in which not only technical models are taken into account, but also socio-behavioural ones. In this work, a number of different models for the parts of an electrical system are presented, related to production, demand and storage. The models are intended to be as simple as possible in order to be simulated in an integrated framework representing the system as a whole. Furthermore, the models allow the inclusion of social behaviour and other, not purely engineering-related aspects of the system, which have to be considered from a complex point of view.

Models that have been created as individual agents to represent specific components can be combined and integrated to represent the energy system. Putting the models together allows the system to be represented by aggregating these models like building blocks in a modular way. In this thesis, the production side is addressed first, through the example of a wind farm. This example allows us to show the behaviour of the wind farm at different time scales and shows the relation of individual wind turbines and the aggregated wind farm. Furthermore, it illustrates the importance of modelling agents in a heterogeneous way, e.g. by parameterising them differently, and including failure behaviour in order to represent the system more realistically. The representation through an agent-based approach is well suited in this case.

After showing the production side, a second example illustrating the demand side is presented. It employs a multi-level model that couples the simulation of individually modelled consumers with a simplified grid model, representing frequency behaviour. Refrigerators were chosen as consumers, because of their availability and thermal storage abilities. Two main findings were made:

The aggregation effect, which describes why and how a load curve flattens when aggregating different number of consumers, was introduced. It was found that the standard deviation is a power law of the number of consumers. This means that for only a few refrigerator agents, fluctuations in the curve are high, whereas for large numbers, the aggregated load curve does not change significantly and its shape is much smoother. This power-law relation is typical for complex systems with many interactions and processes. Even if only tested with the refrigerator model, it can be assumed to hold true for other electrical use and will be investigated further.

The second finding in this work is an emergent phenomenon observed in some of the simulations in which, an under-frequency load shedding (UFLS) was applied. Disconnecting the refrigerators at a frequency drop can improve the stability of the grid, and therefore UFLS mechanisms are used. With a simple UFLS strategy, a rebound effect was observed, which can lead to synchronisation of the working cycles among different refrigerators, which individually have a pulsing load curve. The synchronisation of a large number of them can have fatal effects on the system, as an oscillation of loads and frequency was detected. This phase transition from a stable towards an oscillating system has shown many parallels with synchronisation effects, that have been thoroughly studied in complexity science, such as hands clapping or firefly lightning. The model allowed us to understand and analyse the origins of this phenomena. This was possible through the exploration of individual behaviour of the agents, as well as by using statistical methods such as Monte-Carlo simulations to analyse the behaviour of this non-deterministic model over several cases.

It can be said that, overall, complex systems can be very helpful in approaching the electrical energy system from an integral point of view, where interactions among different scales and levels of the system are important. This is exactly the case in upcoming, distributed and communicating energy systems, or *smart grids*. Using techniques like agent-based modelling as simulation tools for complex systems, allows for simulating effects like emergent phenomena, which can have important effects on current and future systems. These effects are not restricted only to oscillations and synchronisations, as illustrated by cascade effects during blackouts. The complex system approach opens new research topics that are proposed in the outlook of this thesis.

We can conclude in this thesis that complexity theory can help in the improvement and design of future energy grids, as well as in acquiring a better understanding of the operation of the system itself from an interdisciplinary perspective.

Síntesis

Debido al cambio climático y el calentamiento global, así como la preservación del medio ambiente, nuestro mundo se enfrenta a un importante desafío, que exige un uso más sostenible y óptimo de los recursos naturales. La energía es uno de los principales contribuyentes a las emisiones y por tanto existe un gran potencial de mejora en este campo. Una de las consecuencias de la necesidad de sostenibilidad es la tendencia general hacia la introducción de fuentes de energía renovables en los países industrializados. Esto implica uno de los mayores cambios en la estructura de los sistemas energéticos, alejándose de un sistema energético centralizado y jerárquico hacia un nuevo sistema en el que diversos actores, a distintas escalas, influyen en su comportamiento. Además, decisiones políticas como la liberalización del sector, han creado lo que puede llamarse un cambio de paradigma en el sistema energético.

Los sistemas de energía eléctrica suelen ser de gran escala, estar interconectados, compuestos por redes a distintos niveles, y con gran extensión geográfica. Además, el sistema électrico muestra un comportamiento dinámico en el tiempo e integra muchos participantes con diferentes intereses y objetivos locales, que interactúan entre ellos y con el sistema. Basándose en estas suposiciones, podemos asumir que se trata de un verdadero *sistema complejo*. La ciencia de la complejidad es un campo de investigación emergente e interdisciplinario, que ha sido investigado principalmente en los ámbitos biológico, social y físico. La complejidad, como tal, es objeto de investigación no sólo en éstos ámbitos. Su objetivo es comprender mejor y analizar los procesos, tanto en los sistemas naturales como en los creados por el hombre que están compuestos de muchas entidades que interactúan, en diferentes niveles o escalas. Esto puede llevar a comportamientos colectivos a nivel de sistema, que no pueden deducirse directamente de la conducta de los individuos; como por ejemplo, bandadas de pájaros volando en formación, o caminos de hormigas. La ciencia de la complejidad también está estrechamente relacionado con la teoría de redes, que se utiliza para describir las relaciones o interacciones entre las entidades. Así, por ejemplo, la topología de un sistema eléctrico es de tipo *libre de escala* (o *scale-free*), lo que significa que hay muchos nodos con pocas conexiones, pero sólo algunos con una gran cantidad de conexiones. Otra de las aplicaciones de la teoria de la complejidad en relación con los sistemas électricos es el estudio de su *resiliencia* (que describe la elasticidad o resistencia de

(a) Sistema eléctrico clásico

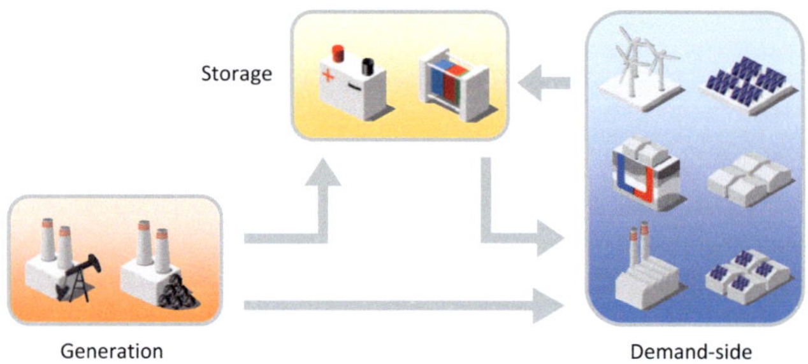

(b) Sistema eléctrico moderno

Figure 1: Cambio de paradigma en los sistemas energéticos eléctricos: (a) muestra los flujos unidireccionales de una red clásica, mientras que en (b) se puede observar un sistema moderno que incluye productores en el lado de la demanda, al igual que almacenamiento de energia. En éste ultimo, los flujos pueden ser en varios sentidos.

un sistema) frente a ataques o sabotajes localizados en su red, en relación a su topología. Sin embargo, en el ámbito de la energía las aplicacions de la teoria de la complejidad aun siguen siendo marginales. En esta tesis, intentaremos recalcar y demostrar la importancia de un enfoque desde el punto de vista desde la cienca de la complejidad, ya que aparentemente hay muchos paralelismos con otros sistemas estudiados ya por esta ciencia. La hipótesis de cómo este enfoque podría ayudar a entender mejor el comportamiento de los sistemas de energía se trata desde un punto de vista teórico, en primer lugar. En una segunda etapa, la aplicación del enfoque se ilustra a través de algunos ejemplos de modelización y simulación.

Una de las maneras de estudiar los sistemas complejos es a través de su modelización y simulación, la cual se utiliza como una herramienta para recrear estos sistemas en un entorno virtual. Los avances actuales en potencia de cálculo computacional (que ha sido una limitación importante en este campo hace algunas décadas), permiten la simulación de este tipo de sistemas dentro de márgenes de tiempo razonables. Una de las herramientas para la simulación de sistemas complejos son los modelos basados en agentes. Este tipo de modelo, centrado en los individuos del sistema, se basa en entidades autónomas que pueden interactuar entre sí. Eso representa el sistema de una manera más natural, de forma desagregada. Los modelos basados en agentes pueden ser acoplados a otros métodos de

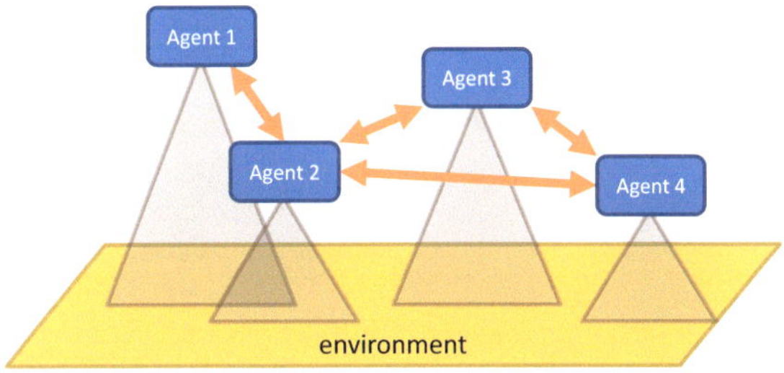

Figure 2: Modelo basado en agentes: los agentes siendo entidades autónomas e independientes, interactúan entre sí y tienen una visión limitada de su entorno.

modelado, tales como los modelos continuos y eventos discretos, que se pueden empotrar en los agentes; o ejecutar en paralelo al sistema multi-agente. Para representar el sistema de energía eléctrica de forma sistémica y en varias capas, éste tiene que ser considerado como un verdadero sistema socio-técnico, en el que no sólo los modelos técnicos son tomados en cuenta. Se han creado una serie de modelos diferentes de las partes de un sistema eléctrico, incluyendo la producción, la demanda y el almacenamiento. Los modelos son desarollados de la manera más simple posible, pero aun así cumpliendo sus objetivos definidos. Esto permite simular un sistema compuesto de gran numero de modelos individuales en un marco integrado, que representa el sistema de manera integral. Por otra parte, los modelos permiten incluir comportamientos sociales y otros aspectos no puramente técnicos, los cuales también influyen o interactúan con el sistema. Estos aspectos interdisciplinarios solo pueden ser considerados desde un punto de vista complejo, ya que tratamos con modelos de ámbitos muy distintos, pero aun asi relacionados entre sí. Esto puede permitir la representación de efectos complejos emergentes en el sistema eléctrico tal y como sincronizaciones o efectos en cascada, que tienen mucho parecido con los fenomenos emergentes estudiados ya en la teoria de la complejidad aplicada a otros campos.

Para representar el sistema eléctrico, se han creado modelos de sus componentes. Estos han sido diseñados como agentes individuales y pueden ser combinados e integrados de distintas maneras, en función de la cuestión a tratar. Esta manera de representar las distintas entidades del sistema permite combinarlas de una forma modular y flexible, permitiendo la reutilización de los mismo en distintos modelos sistémicos. En esta tesis, en primer lugar se enfoca el lado de la producción, mediante la creación de un modelo de un parque eólico. Éste modelo permite mostrar el comportamiento del parque para diferentes escalas de tiempo y muestra la relación entre los aerogeneradores individuales y el parque eólico agregado. Además, se ilustra la importancia de la heterogeneidad entre agentes, la cual aporta más realismo al modelo. Esto se consigue por ejemplo usando parametrizaciones ligeramente diferentes para cada agente, ya que en la realidad tampoco los generadores son completamente idénticos. Adicionalmente, se incluye el comportamiento ante la conexión

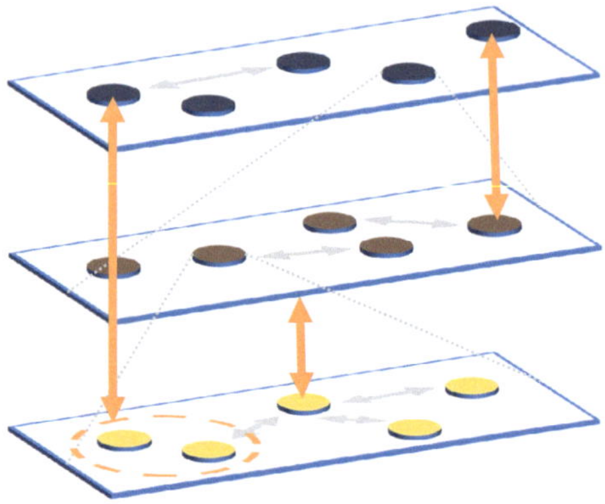

Figure 3: Interacciones entre distintos niveles del sistema. En gris, las interacciones dentro de una misma capa, que son tomadas en cuenta en modelos clásicos. En naranja, las interacciones entre distintos niveles, que son una de las bases de la complejidad de un sistema, como por ejemplo el eléctrico. Las acciones de los consumidores individuales pueden tener efectos a otros niveles totalmente distintos del sistema, como la frecuencia de la red emerge la agregación de todas sus entidades. De nuevo, este valor puede tener un efecto en los consumidores individuales, con lo cual tenemos un bucle dinamico que conforma un sistema no lineal.

y desconexión de aerogeneradores de forma individual con el fin de representar las bajas por mantenimiento y fallos de los aerogeneradores. La representación a través de un modelo basado en agentes es muy adecuada aquí.

Habiendo mostrado el interés desde el lado de la producción, se presenta un segundo ejemplo desde el lado de la demanda. Se trata de un modelo multi-nivel que conecta la simulación de consumidores individuales con un modelo simplificado de la red, el cual es usado para representar el comportamiento de la frecuencia de la red. Como aparatos consumidores, se escogieron electrodomésticos frigoríficos, debido a su gran propagación y su capacidad de almacenamiento térmico. Las dos conclusiones más importantes del modelo se describen a continuación.

En un primer paso, se presenta el efecto de agregación, que describe cómo y por qué una curva de carga se aplana, al agregar la carga de distintas cantidades de consumidores. Se encontró que la fluctuación, descrita por la desviación típica de la curva de carga, es una ley de potencia del número de consumidores. Esto significa que para sólo unos pocos frigoríficos, las fluctuaciones en la curva son altas, mientras que para un gran número, la potencia no varia y la curva de carga total tiene una forma mucho menos abrupta. La relación de ley de potencias es típica en sistemas complejos con muchas interacciones y procesos. Aúnque solamente se ha mostrado con un modelo relativamente simple como el de unos frigoríficos conectados a un modelo de red simplificado, una generalización a otros casos y aparatos eléctricos parece razonable.

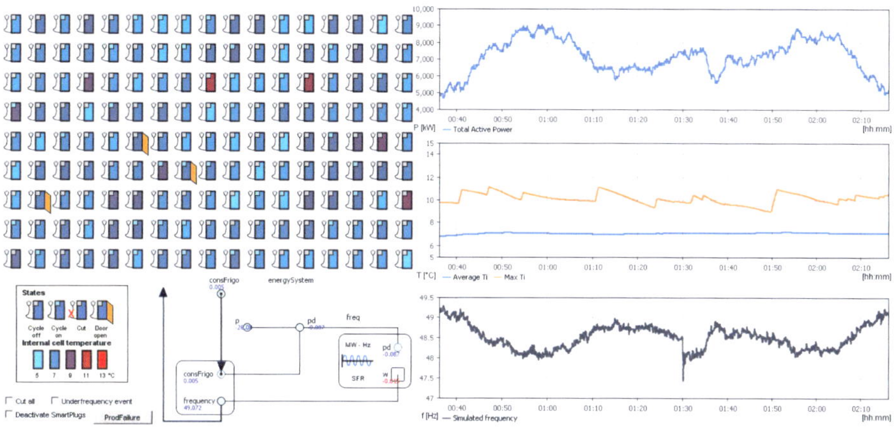

Figure 4: Entorno de simulación para la gestión activa de la demanda. La poblaciíon de frigoríficos representados individualmente es conectada a un modelo de la red que representa su frecuencia. Esto permite simular escenarios de gestion basada en frecuencia, al desconectar los electrodomésticos al detectar una caida anormal. Tal y como muestra la imagen anterior, en este ejemplo se acoplan modelos a distintos niveles (el nivel de los consumidores individuales por un lado, y el nivel sistémico, representando el sitema eléctrico con su generación y demanda, por el otro). Este modelo multiescala integrado permite analizar el comportamiento dinámico y las interacciones entre carga y frequencia de este sistema complejo, y observar fenómenos emergentes.

El segundo hallazgo fue un fenómeno emergente que puede ser observado en algunas de las simulaciones, en las que se implementó la técnica de desconexión por caida de frecuencia (under frequency load shedding, UFLS). Desconectar los frigoríficos durante una caída de frecuencia puede ayudar a mejorar la estabilidad de la red, para lo que se utilizan mécanismos del tipo UFLS. Con una estrategia UFLS sencilla, se pudo observar un efecto rebote, el cual puede llevar a una sincronización de los ciclos de carga de los diferentes frigoríficos. Estos, individualmente, tienen una curva de carga basada en pulsos cíclicos, pero de características y duraciones distintas entre ellos, dependiendo del tipo y uso del aparato. Una sincronización de estos ciclos puede tener efectos nefastos en el sistema, como una oscilación de cargas y frecuencia. La transición de fase de un sistema estable a un sistema oscilante ha mostrado muchos paralelismos con efectos de sincronización que han sido estudiados anteriormente por la ciencia de la complejidad. Algunos ejemplos son la sincronización de aplausos del público en un espectáculo, o la sincronización de las señales luminosas de las luciérnagas, cuyos periodos de luminosidad intermitente se sincronizan de manera emergente en la naturaleza. El modelo nos ha permitido comprender y analizar los orígenes de este fenómeno. Esto fue posible a través de la exploración del comportamiento individual de los agentes, así como mediante métodos estadísticos, tales como simulaciones de Monte Carlo para analizar el comportamiento de este modelo no determinista para un gran numero de casos distintos.

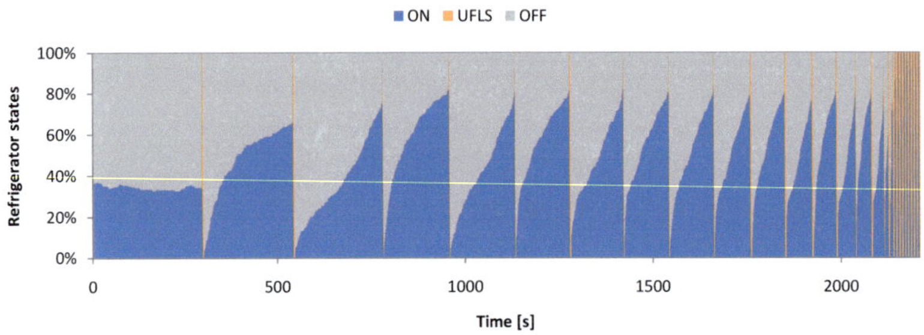

(a) Sincronización de una población de frigoríficos

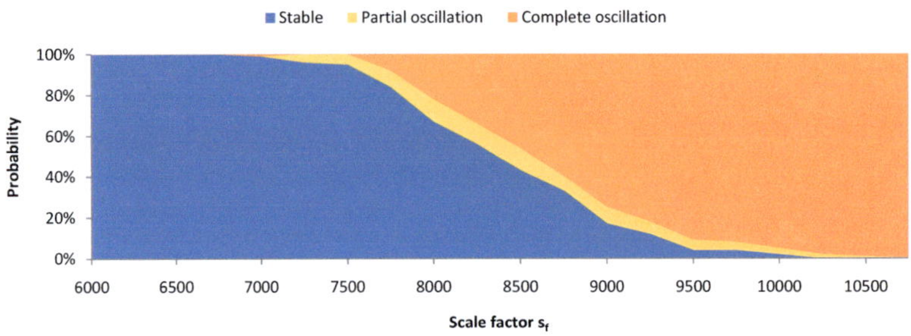

(b) Cambio de fase de un sistema estable a uno oscilante

Figure 5: Sincronización y oscilación de una población de frigoríficos: Usando un sistema de gestion de carga basado en frecuencia (UFLS) se pueden observar sincronozaciones emergentes (a). La aparición de estas oscilaciones se cuantificó mediante una probabilidad al tratarse de simulaciones no deterministas. La probabilidad se analizó variando el grado de influencia de los frigoríficos en el sitema eléctrico global. En (b) se puede observar un cambio de fase, que sigue una curva en forma de «S», típica en sistemas complejos.

En general se puede decir que un enfoque desde los sistemas complejos puede ser muy útil para representar el sistema eléctrico desde un punto de vista integral, donde las interacciones entre diferentes escalas y niveles del sistema son fundamentales para la comprensión de causas y efectos. Este tipo de interacciones serán de especial importancia en el caso de sistemas distribuidos y comunicantes, tal y como será el caso en la próxima generacion de sistemas eléctricos, las *smart grid* o redes inteligentes. El uso de técnicas para simular sistemas complejos, como por ejemplo la modelazación basada en agentes, permiten la simulación de efectos como los fenómenos emergentes, que pueden tener efectos importantes en los sistemas actuales y futuros. Estos efectos no se limitan a oscilaciones y sincronizaciones solamente, sino que tambien engloban otros fenómenos como por ejemplo los efectos en cascada como muestran algunos apagones eléctricos. El enfoque desde el punto de vista de la complejidad abre nuevas lineas de investigación que son propuestas en las perspectivas finales de esta tesis. Podemos concluir que la teoría de la complejidad puede ayudar a la mejora y el diseño de los sistemas électricos venideros, así como la adquisición de un mejor conocimiento del funcionamiento del propio sistema, desde un punto de vista interdisciplinario.

Acknowledgements

This thesis would not have been possible without the support of many people. First and foremost I offer my sincerest gratitude to my supervisor, Oscar Barambones, who has supported and guided me throughout this time, with his help and advice, whilst also allowing me the room to work in my own way. My sincere thanks also goes to José María González de Durana, for his patience and valuable knowledge which contributed to this work, as well as for his continuous support in almost any aspect; and for being always available for very fruitful exchanges and discussions.

I wish to especially thank Pablo Viejo, who introduced me to this fascinating topic almost five years ago already, and without his support this thesis would not have been possible. He has made available this support in a number of ways, through always inspiring and motivating discussions, valuable assistance and suggestions.

I would like to express my gratitude towards my numerous colleagues at EIFER and EDF, for their daily motivation and aid. I especially thank Fabrice Casciani, Kevin McKoen and Patrice Nogues for offering me the possibility to contribute to their projects through this work. I also would like to thank Carolina Tranchita for her comments on the manuscript. Furthermore, I would like to thank André Lachaud for his valuable support with the refrigerator measurements.

Also I would like to thank the colleagues from SIANI - University of Las Palmas, especially Jose Juan Hernandez and Mario Hernandez for the fruitful discussions. I would like to show my gratitude to all the students, among them Alvaro, David, Nicola, Jennifer, Izaskun, Jose, Irantzu, Susana, Thibault, Sergio and Johannes, who worked with me throughout this time, and contributed in one way or another to this thesis. Besides I would like to thank Caroline McKoen for her patience while proofreading, which not only contributed to improve grammar and spelling, but also served to clarify many ideas through intense discussions. I also thank my friends who supported me during this time.

Finally, I thank my parents and sister for supporting me throughout all my studies and this thesis.

Enrique Kremers, May 2012.

This thesis was possible through the support of the European Institute for Energy Research, EIFER (EDF & KIT), which gave me the opportunity to do this work in a multi-disciplinary atmosphere. I would explicitly like to acknowledge the financial, academic and technical support of EIFER and EDF.

Contents

Chapter 1

Introduction

Our society is facing major challenges in achieving and maintaining a sustainable and environmentally friendly energy supply. The demand for energy is constantly growing, and electrical energy, especially, has escalated in importance during the last century with a continuous increase in electrical devices. Independent of efforts to reduce consumption and improve efficiency, the growing use and quantity of electrical equipment has contributed to, and encouraged, a rise in energy demand.

In parallel with this increase in consumption, global issues have arisen which do not allow the uncontrolled or unlimited expansion of power generation, needed to cover demand. Fossil fuels are limited and impose a natural constraint, however their use during the last centuries has had a negative impact on the environment. Global warming, which is partially due to the greenhouse effect and is related to increased CO_2 emissions (which are also produced by burning fossil resources), further constrains the expansion of the classical generation system. In addition to this, fossil fuels also contribute to the pollution levels through inefficient and unsustainable energy generation.

These issues are forcing policy makers to find solutions and improve the sustainability of the energy supply system. At the same time, they must guarantee it is both economically affordable and technically secure. Thus, decision makers, often at governmental level, have an influence on energy stakeholders.

Since the deregulation of the energy markets, which began in the late nineties and has been promoted by the European Union, the number of actors has increased considerably. Former state-owned utilities have been transformed into profit making companies. These utilities and operators along with and other actors are constrained by the European Commission's legislative framework and by individual national initiatives. Their drive to achieve economic benefit is limited by this legal framework which, for example, currently proposes to reduce CO_2 emissions, by the introduction of emission certificates and incentives promoting renewable and sustainable energy sources.

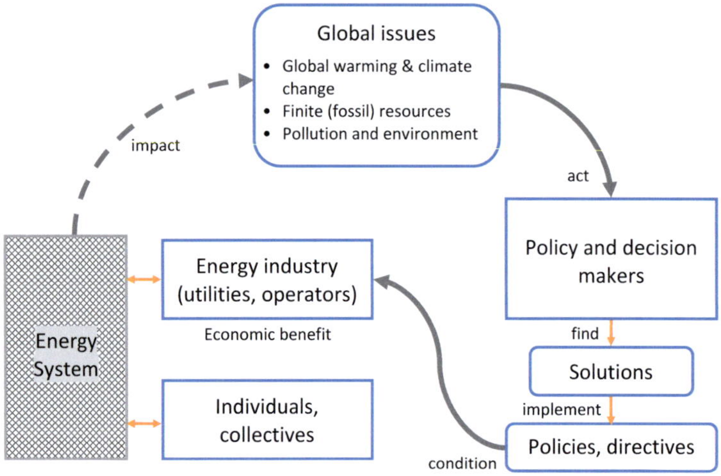

Figure 1.1: Global framework of the energy system.

This is leading to a paradigm shift in the electrical system. We are at a turning point, mainly because of the deregulation of the sector, environmental constraints and because of the introduction of renewable and distributed energy sources. Based on these trends, buzzwords such as the *smart grid* are common. Considering the future system as a smart electricity system (whatever might be meant by *smart* at this point), it can be said that there is a movement towards a modified system which is challenging the energy industry as well as the current power system itself.

What the future system will look like is difficult to predict. However, given the framework above it can be said that it will have to be able to achieve a number of objectives. Possibly, its rather hierarchical, unidirectional structure will be largely affected through an increased number of distributed resources being integrated. Furthermore, the use of renewable energy sources will increase. The significant and sometimes unpredictable fluctuation of renewable energy production means that these sources cannot be controlled in the same way as classical production stations. An increase in supply from renewable sources results in the need to develop new systems for production control and a growing need for effective and flexible generation units to compensate for this variability in supply.

There are many solutions and ways towards the smart electricity system and these depend on many factors. Location and local conditions may affect this process, as well as different national or regional policies. The much discussed question of the *optimal* energy mix or implementation of concrete technologies such as smart metering or electric vehicles are only some examples.

In order to address these and other issues, modelling and simulation can support and facilitate the transition towards a smarter electrical system. Simulation is a procedure in which (usually dynamic) systems can be analysed through running experiments on a model specifically designed for this purpose. Exploring different ways of tackling the challenges of the future system through simulation supports decision-making on complex questions.

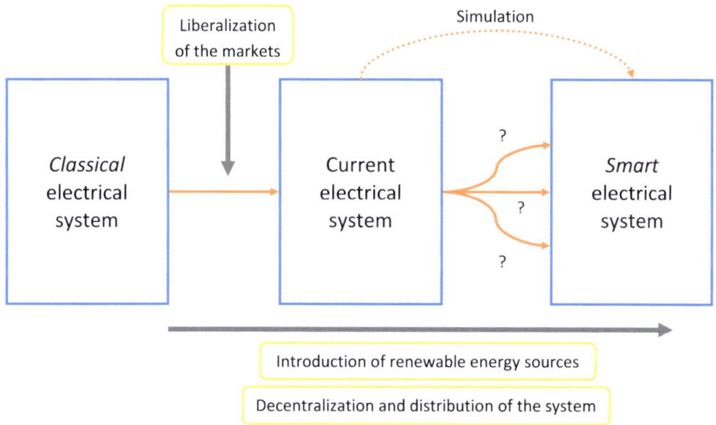

Figure 1.2: Evolution of the electrial system.

Many different tools have been used for simulating electrical systems. However, the paradigm shift towards the smart grid now raises the question of whether there is a need for new approaches considering that new issues are, and will be, arising. The classical electrical system (see Figure 1.2) is a rather hierarchical and one-directional system, where production is centralised and injects on the one side, and demand is distributed and consumes on the other. Production only needs to guarantee to meet demand.

This operating principle is no longer valid. The inclusion of distributed production at lower levels of the system has created the possibility of local production, which can invert, or at least reduce, the classical, one directional flow from big stations towards the final end user. As many demonstration projects show, in the future grid it will be possible to manage the demand side. Demand is no longer regarded as a static, unmanageable part of the system, but rather it will support balancing and stability processes. Managing and controlling these processes will require distributed control and communication mechanisms.

In order to correctly represent this system in an integrated, systemic approach, it is first necessary to create a model of the existing physical system. The representation of the classical system will be achieved by taking into account both current and future possibilities; this will allow for it to be continually extended. Completing the model with current technologies such as smart metering, distributed and renewable generation, etc. will allow for the representation of the state-of-the-art.

The inclusion of a communication layer already seems reasonable; it is needed to handle control and management processes over the distributed entities of the system. The current system model can be modified for possible future scenarios by including prospective technologies and implementations. The analysis of this virtual system will allow us to extrapolate and identify future challenges. The model will enable the simulation of concrete case studies relating to real-world sets of problems in the transition to a smart power system.

The detailed knowledge required to accurately model a system is challenging to gather. This thesis explores modeling the *electricity system* however, this term is not simply defined. The different networks in different countries, each made with different technologies create a complex overall system. So how can we make a model of a system that is not fully known?

The solution adopted in this thesis uses a design that models the system in a simplified way, rather than initially focusing on details. The aim is to create a systemic model, rather than detailed models of a part of it. Furthermore, the objective is to create an individual-centric rather than a system level model. This means that individual parts of the system will be represented as such, and not by aggregated models.

This will help to define a new, hypothetical smart grid. The model of the energy system, has been designed for the minimum structure and components needed to emulate the essential behaviour of the current and future grids. It will serve to make virtual experiments (simulations) of the current energy system and also of potential future scenarios which are not yet possible in current networks, or which would require great expense or risk.

In the first part of this thesis, complex system theory is presented. There are only few examples in which this relatively new field of research has been applied to energy systems. As we will see however, there is great potential to describe an energy system through its complexity. Therefore, we will analyse some of the main aspects of an electrical energy system relevant to this task. Afterwards, the question of why we should treat this system as a complex one is discussed.

In the third chapter, the subject of modelling and simulation of complex systems is addressed. Varied approaches to this field are presented as well as a combination of different methods. This serves as a basis for the models which follow.

The chosen approach, along with its related challenges, is presented in chapter four. The tools and models used are shown, for both the production and demand sides, as well as the network itself. In this way, a minimal energy system with all of its basic components is represented.

The fifth chapter presents two case studies and demonstrates how they are integrated using the models previously described. The first case study addresses the production of renewable energy by representing a wind farm. It is based on individual and heterogeneous wind turbine models. The second case study proposes a more systemic model, which includes refrigerator appliances as consumers, and a grid model to represent the complete system behaviour.

In the last chapter, the simulation results of the two case studies are presented and discussed. Some relevant findings are included to illustrate the advantages of using the selected approach. Scale effects, as well as interactions and emergent phenomena found in the simulation, are discussed. Finally, the conclusions drawn from these studies are presented, as well as a look at future possibilities for research.

Chapter 2

Complex Systems in the Context of Energy

The whole is more than the sum of its parts.

Aristotle, around 350 BC.

Complex systems are characterised not only by a large number of components, but also by the diversity of these components, their relationships and interactions. In these systems, the relationship between the components can lead to collective behaviour at a system level. This is a subject of research in what is called *complexity science*. In the 1940's and 1950's initial approaches were made towards systems theory and artificial intelligence. During the 1960's and 1970's, formulations on what is now known as complexity science were made on phenomena on order patterns in market systems. A phenomenon is emergent in the sense that it results from human actions, but is not of human design [Hayek, 1967]. Later on, other fields such as network theory, game theory, agent-based modelling, fractals and chaos theory where found to be closely related to complexity science. Since then, this interdisciplinary field has been a focus of research into different domains, mainly statistical physics as well as in social, biological and computer sciences, sometimes with quite diverse scopes. Some examples on current application of complex systems theory are shown in Figure 2.1 and are classified into different disciplines.

Energy systems are composed of many components at different levels connected to each other, at a physical level, through a network infrastructure. The paradigm shift in the energy sector characterised by the deregulation of markets, the introduction of renewable energy sources and system decentralisation, has increased its degree of complexity. For instance, the introduction of *smart grids* will result in new communication means used within the system. This will make an already large system much more dynamic in terms of interactions and communications. Therefore, we will see in the course of this chapter, that

Figure 2.1: Some of the disciplines which are related to complex systems.

a complex system approach can be helpful. This is an emerging field of research [Miorandi et al., 2010].

The first part of the chapter presents the theory of complex systems as well as the properties and some examples of them. The second part gives an introduction to modern energy systems. Finally, the characterisation of energy systems as complex systems is discussed.

2.1 Defintion: What are Complex Systems?

Complex systems are studied by complexity science, an interdisciplinary field that considers all kind of systems which are constituted by many parts with numberous interactions, and investigates their behaviour.

To explain phenomena concerning complex systems, their close relationship with networks, swarm intelligence, game theory and mathematical and stochastic systems must be considered. Regarding complexity from a multi-disciplinary perspective can help us to understand the behaviour of these systems. A common definition for complexity science remains an open question, as is appears in many different disciplines and encompasses many different views. These perspectives reach from complex mathematical systems to social sciences, passing by biological and technological systems. Despite conceptual advances in concrete fields like chaos theory or emergence in non-linear or self-organised system, which were studied in the previous decades, a complete theory of complexity does not yet exist [Barabási, 2007].

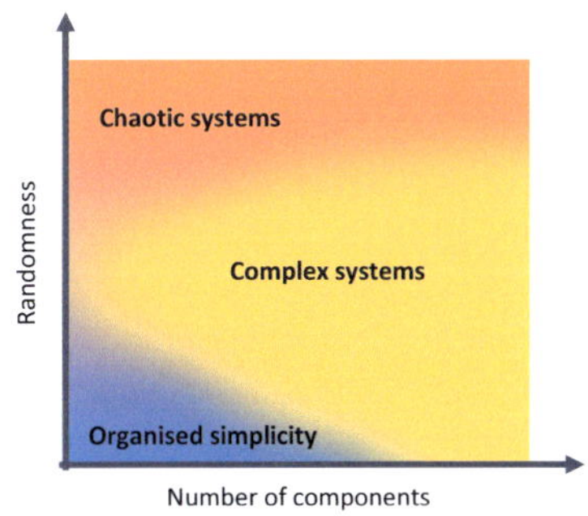

Figure 2.2: Organization and randomness in systems: Complex systems are located somewhere between order and chaos, and are usually made up of a large number of components.

The following definitions provide an overview of the range of understanding of complex systems. According to different authors, a complex system might be:

- "A complex system is comprised of a (usually large) number of (usually strongly) interacting entities, processes or agents, the understanding of which requires the development, or the use of, new scientific tools, nonlinear models, out-of equilibrium descriptions and computer simulations", [ACS, 2010];
- "A system that can be analysed into many components having relatively many relations among them, so that the behavior of each component depends on the behaviour of others" [Simon, 1996]

- "A system that involves numerous interacting agents whose aggregate behaviours are to be understood. Such aggregate activity is nonlinear, hence it cannot simply be derived from summation of individual components behaviour." [Singer, 1995]

Simon's definition of complex systems is therefore, rather, a description in the structural sense, whereas Singer's definition is concerned more with the dynamics or functional relations of the system. Furthermore, the concept of organisation or structural order plays an important role: "Complexity is neither simple order nor a complete mess. It is something between order and chaos, and it grows at the edge of chaos. A complete mess or chaos cannot be represented in any shorter or more compact way than the mess itself. A simple and static order on the contrary can be represented as a short formula. Complexity is different from both of these, and although it often is a result of rather simple formulas too, it includes iterations, the repetition of patterns - taking part of the result of the former round as the input to the next - and most often also adding some randomness to the process. This means that complexity is a result of a process unfolded in time." [Grönlund, 1993]

A historical overview of complexity science is shown in Appendix A.

2.2 Properties and Features of Complex Systems

2.2.1 Emergence

One of the best known quotations related to complex is from a famous ancient Greek philosopher: "The whole is more than the sum of its parts" [Aristotle, around 350 BC]. Complex systems often behave in unexpected ways that cannot directly be inferred from the behaviour of their components; this is known as emergent behaviour. This is one of the most important particularities when regarding a system from a complex point of view. In classical approaches the system is usually deterministic or at least predictable (stochastically) in a large sense. Emergence describes how complex behaviours at the macro-level of a system arise out of more simple ones. These behaviours entail emergence of properties that can hardly, if at all, be inferred from the properties of the individual parts.

Aristotle doesn't suggest that complex systems are completely unpredictable. Complex systems behave in a way that has to be studied, taking into account these particularities, in order to better understand their operation and dynamics. In real systems, emergence can either be a desired or undesired effect. If, in a real system, emergence is found, it can be influenced to avoid unwanted effects and can make the system evolve in the desired direction. Therefore, a better understanding of emergent phenomena can help us in designing or managing those systems.

In the following, some examples of emergence in different fields are shown.

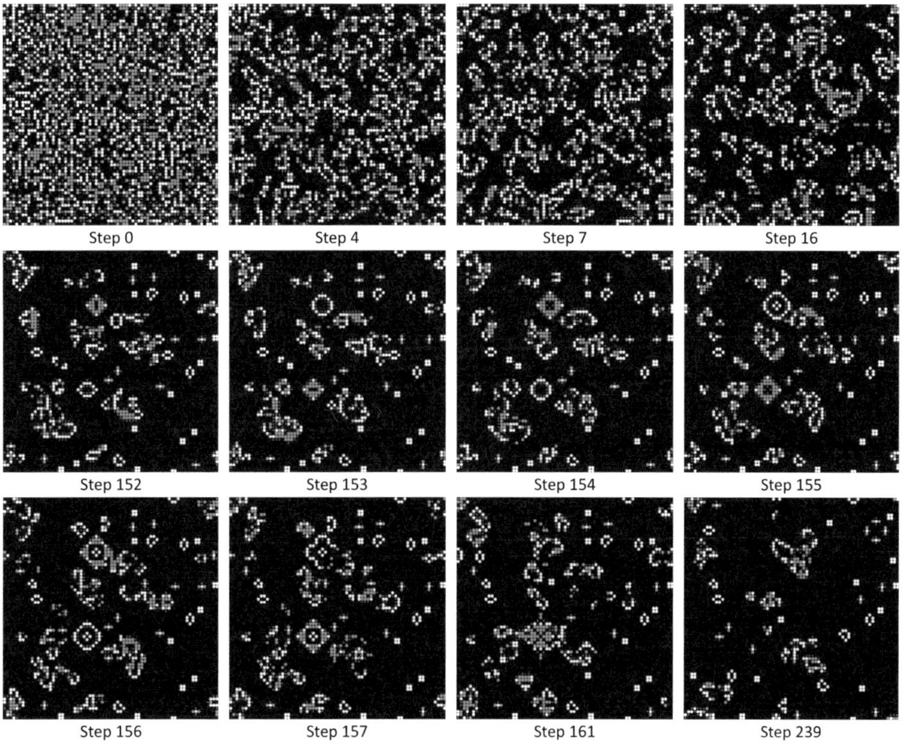

Figure 2.3: Sequence of some generations of the *Game of Life*. Starting from a random setup (step 0) it can be seen how the system evolves into emergent structures, such as symmetrical, pulsating or moving patterns (steps 152-157), which cannot be inferred by the rules of the individual cells. The sequence was created using NetLogo [Wilensky, 2005]

Mathematics: Cellular Automata

Conaway's *Game of Life* is a two-dimensional cellular automaton, whose cells can either be alive or dead. Simple rules for each cell determine their change of state (dead/alive). Depending on the initial conditions, the whole system of cells can show an extraordinary complex, ordered behaviour, which does not allow the inference of the primitive rules of the individual cells. These rules are simple, for example that a new cell is born if the number of living neighbours is in a specified range, or that it dies when there are too many or too few living neighbours [Packard and Wolfram, 1985]. The Game of Life is one, and probably the best known, of several cellular automata. Changing rules and dimensions allows for the creation of other automata of any complexity [Von Neumann and Burks, 1966].

Physics: Percolation

The phenomena of percolation describes the behaviour of subgraphs in a random graph[1] and how they form components or clusters.

(a) $p << p_C$ **(b)** $p < p_C$ **(c)** $p = p_C$

Figure 2.4: Percolation of oil in a porous soil showed using a Netlogo model [Wilensky, 1998, 1999]. The probability p represents the porosity of the soil. With increasing p, it is more likely that the oil reaches the bottom. However, the percolation probability $\theta(p)$ is not linear, but rather a phase shift. From a given threshold onwards, it is very probable that the oil will reach the bottom - but before this threshold, it is very unlikely.

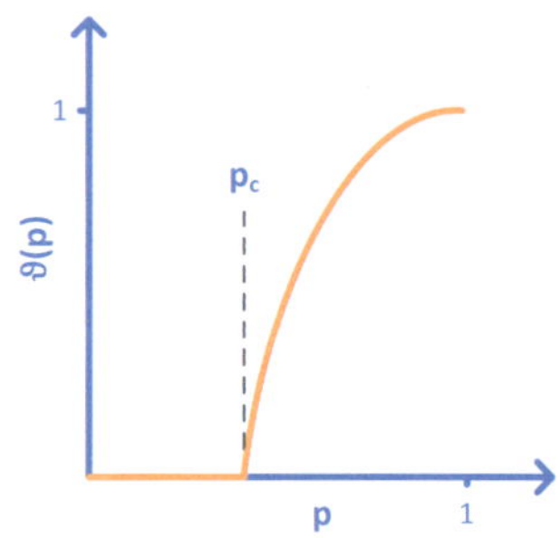

Figure 2.5: $\theta(p)$, the probability that there is a uninterrupted path from the top to the bottom of the graph, indicates that above the critical probability p_c there is a high probability of percolation. The function θ is continuous except possibly at p_c [Grimmett, 1999]

Percolation models can be used to represent the flow of fluid in a porous medium with randomly blocked channels [Bollobás and Riordan, 2006]. The effect is described in the

[1]A random graph is generated by a random process, such as adding edges with a probability p (see Section 2.2.2).

following way: if for a given probability p (which describes whether an an edge between two nodes exists and permits passage) there is an open path from the top to the bottom of the network, and at which probability this path would exist. There is a critical value p_c at which the probability of the existence of a path increases sharply, this means that below p_c there is no path, but it is likely that above this value such a path exists, as can be seen in Figure 2.4. This almost *discrete* effect, which happens suddenly, cannot be inferred from a continuously growing probability p without taking into account the complexity of random networks; this is another example of emergence. The percolation function is also related to power laws, another typical relationship appearing in complex system [Austin, 2011].

Biology: Honeybee Hives

In bee colonies, the phenomenon of *thermoregulation* can be observed. The brood nest needs to maintain the same temperature over a long period to develop the brood; this temperature is also optimal for the creation of wax. A complex mechanism involving large numbers of bees is used to maintain a stable temperature inside the hive [Jones et al., 2004]. The individual bees have no knowledge of the overall hive conditions, and only sense their immediate environment. Each bee has different methods to regulate its temperature, such as wing fanning, building isolating wax layers, evaporative cooling, etc. The thermoregulatory mechanism of the hive is self-organised and arises from simple rules followed by each bee. In very extreme conditions, even bee corpses fulfil an insulating function.

Computer Science & Biology: Swarms

Swarming or flocking is a typical example of a collective behaviour in which emergence arises. Swarming in nature is a collective animal behaviour exhibited by a large number of insects or other animals. As a swarm is not a static phenomenon, the individuals adapt their direction of movement and have to avoid collision. This creates the swarm, a collection of individual animals or units moving together as a group.

A typical swarm can be seen in Figure 2.6. The emergence of swarming has been an object of research and modelled and simulated in order to understand, at macro levels, the underlying rules which govern swarms. A swarm can be simulated by modelling individual entities maintaining a certain distance from each other, among other rules.

Architecture & Engineering: The Millennium Bridge

The Millennium bridge was opened in London in the year 2000 and was intended to be "a pure expression of engineering structure" by its designers. However, as thousands of pedestrians began walking over it, the bridge began to sway slightly. Then, suddenly, the bridge began to oscillate and the people began to walk side-ways, but in a perfectly synchronised manner. The oscillating bridge and the walking patterns of the pedestrians caused a positive feedback, each one reinforcing the other. The bridge had to be closed

(a) Flock of birds **(b)** Flock of *boids*

Figure 2.6: Swarming: On the left, a flock is formed by the behavioural rules of individual birds and changes dynamically over space through time. Even so, the flock keeps a coherent and organised structure when zooming out from the individual bird level. On the right we can see a flock simulation. Each bird (or *boid*, from the name of an artificial life simulation program [Reynolds, 1987]) is able to see the other birds and obstacles in a fixed radius in front of it. The bird adjusts its velocity based on a set of simple rules. This implementation also features trees (white circles), which the birds will avoid flying into. The birds form tight flocks and navigate around obstacles as an emergent behaviour. Source: Occoids simulation by Cosmos Research, `http://www.cosmos-research.org/demos/occoids/`

immediately, and redesigned in order to avoid oscillation. This is an example of a costly design where an emergent phenomenon appeared which had not been taken into account through classical design and calculation methods. It was subject of an article of research by Strogatz et al. [2005].

As we see, emergence can occur in very different areas, related to completely different subjects. However, the principles on which it relies are similar across disciplines, and are one of the main topics of research in complexity science.

2.2.2 Complex Networks

Complex systems are composed of many non-identical components connected through diverse interactions. If we formalise these interactions, we end up with a mathematical graph or, more generally, a network. Complex networks are one of the fields of study in complexity science. They deal with the non-trivial, topological features of simple networks, but which can be observed in reality. These include patterns, which are neither completely regular nor completely random. The study of complex networks is inspired by real networks, mainly computer, social and biological. Studies in relation to energy systems are still marginal, however.complex systems are very relevant to energy systems, which are mainly based on a network infrastructure - the power grid. An overview of this field follows.

Complex networks have been object of several scientific studies. The first approach dates back to 1959, when [Erdős and Rényi, 1959] suggested modelling networks as random graphs. In a random graph, the nodes are connected randomly by a placing a number of links among them. Random graphs allow the representation of generic random networks and permit the study of many properties of real systems. However, these assumptions may not be appropriate for modelling phenomena in the real world.

Watts and Strogatz [Watts and Strogatz, 1998] defined p as the probability of rewiring an edge of a regular lattice graph. They analysed networks with wiring probability values between zero and one and found that these systems can be highly clustered and have small characteristic path lengths. Networks with short average path lengths are considered small world networks. This is the case in many (real) networks with a large number of connections. The Watts-Strogatz model first stated that special small world networks exist and are highly clustered. However they don't have as many connections as regular or random graphs but still have short average path lengths.

However, analysing real-life networks has shown that there are also many examples in which the distribution of the number of connections per node is not bell-shaped. Barabási and Albert [1999] found that many real networks have a common property: a power law distribution, which states that the number of nodes with many links is small whereas an exponential growth exists when moving towards nodes with fewer connections. These networks are called scale-free and they are located in between the range of random and completely regular wired networks.

Networks exist everywhere and at every scale [Barabási, 2007]. The human brain, societies or ecosystems are only a few examples of what can be represented as a complex network. Many systems in the real world fulfil the properties described above: neural networks, social networks and also the power grid. For the power grid, generators, transformers and substations were taken as vertices and high-voltage lines as edges.

Metrics of Complex Networks

A complex network can be represented as a directed or undirected graph. In order to characterise complex networks, different metrics have been defined. To explain the most important metrics, a complex network will be represented as a directed or undirected graph $G = \{V, E\}$, where V is the set of vertices or nodes and E is the set of edges or connections. The total number of vertices is $|V|$, and the total number of edges is $|E|$.

- Degree distribution:

  The degree $d(v)$ of a node or vertex v is the number $|E(v)|$ of edges at v; this is equal to the number of connecting edges or neighbours. The average degree $d(G)$ of a network G is given by

$$d(G) = \frac{1}{|V|} \sum_{v \in V} d(v) \tag{2.1}$$

The degree distribution $P(d)$ of a network is defined to be the fraction of nodes in the network with degree d. The distribution can be plotted as a histogram and shows the number of nodes with a given degree.

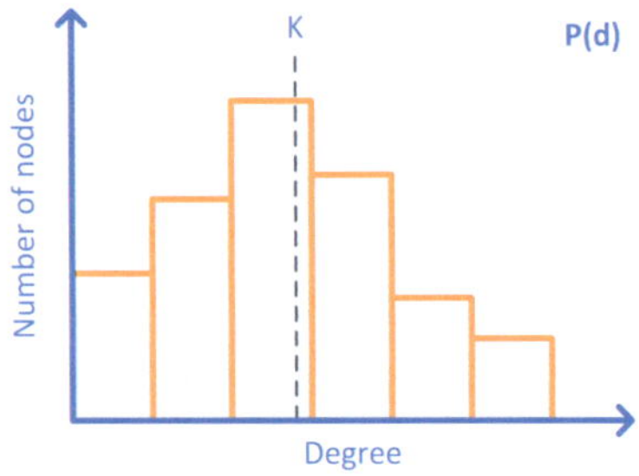

Figure 2.7: Example of a degree distribution $P(d)$ with average degree $d(G) = K$

The average degree is a measure for how many connections a node has on average. A large $d(G)$ implies many connections per node. However, it does not reveal how closely nodes are clustered or how well they are connected to each other (for example, if you wanted to travel around them and find a shortest path, etc.). The distribution of the degrees of the nodes gives us an idea of the spread of connections. If most of the nodes have a similar number of connections this will lead to a peak distribution, whereas if we have heterogeneous numbers of connections among the nodes, exponential or other kinds of distributions emerge.

- Average path length:

  The path length or distance $l(v,w)$ between two nodes $v,w \in V$ is the shortest path between them, measured in number of edges. The average path length $l(G)$ of a graph is calculated over all possible pairs of nodes.

$$l(G) = \frac{2}{|V|(|V|-1)} \sum_{v,w \in V} l(v,w) \tag{2.2}$$

  The network diameter $l_D(G)$ is the maximal path length of the network over all pairs of nodes of G.

$$l_D(G) = \max_{v,w \in V} l(v,w) \tag{2.3}$$

  The path length between to nodes is an essential measure in graph theory, which is fundamental to solving the shortest-path problem, for example for finding the shortest route on a road-map from one point to another. The average path length of a graph indicates how well connected the network is in terms of distances on average. A network with a low average path is likely to be travelled with short distances from any point to another. This plays an important role in small networks, where short average path lengths are the case.

- Clustering coefficient:

  The neighbourhood of a node v are the $d(v)$ nodes that are at distance 1 from A. If v has $d(v)$ neighbours, the number of pairs of those neighbours is

$$h(v) = \frac{d(v) \cdot (d(v) - 1)}{2} \tag{2.4}$$

  The number of pairs which are connected to each other is $f(v)$. The clustering coefficient of node v is then calculated as

$$c(v) = \frac{f(v)}{h(v)} \tag{2.5}$$

  If a node has a high clustering coefficient, this means that it is likely that its neighbours are connected to each other. $c(v)$ can also be interpreted as the probability that two neighbours are connected.

  For a graph G with $|V|$ nodes, we define the clustering coefficient of the network $c(G)$ as

$$c(G) = \frac{1}{|V| \sum_{v \in V} c(v)} \tag{2.6}$$

  If most of the nodes of a network have high clustering coefficients, there are probably many edges connecting nodes together. The network clustering coefficient is a metric that describes if there are many well connected components (clusters) in the graph. However, it does not tell how well these clusters are connected to each other, as a graph with isolated components may have a large clustering coefficient, even without *shortcuts* from one to another cluster.

Types of Complex Networks

When analysing complex networks in the real world, recurrent types of networks were identified among different disciplines and fields. Creating a general classification of networks allows us to find common properties, problems and solutions among these networks, and treat them independently of their original field. In the following section, the most typical networks are presented; these have been identified by different scientists and are generally accepted. The network types can be characterized by the metrics defined above, as well as by other, additional properties.

- Fully connected networks

  Fully connected networks are homogeneous networks in which every node is connected to all the other nodes. If $F = \{V, E\}$ is a fully connected network, the maximal path length $l_D(F) = l(F) = 1$ is equal to the average path length; i.e. every node can be reached from another in only one step. However, the number of connections

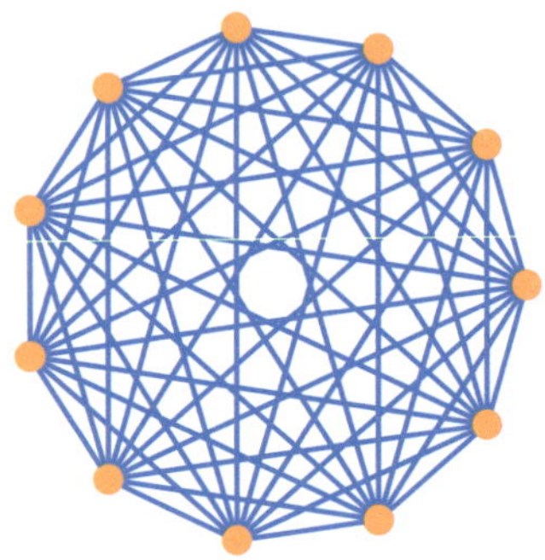

Figure 2.8: Fully connected network

grows rapidly, as a quadratic function of the number of nodes. The clustering coefficient is also $c(F) = 1$, which denotes the most highly clustered network possible.

A fully connected network has the maximum number of edges m, which equals

$$m = \frac{|V|^2 - |V|}{2} \tag{2.7}$$

- Distance based networks

  In a distance based or (regular) lattice network, nodes are connected only to other nodes within a certain range. A ring lattice network, in which the circular layout offers a good visualisation, is a good example of this.

- Ring lattice networks

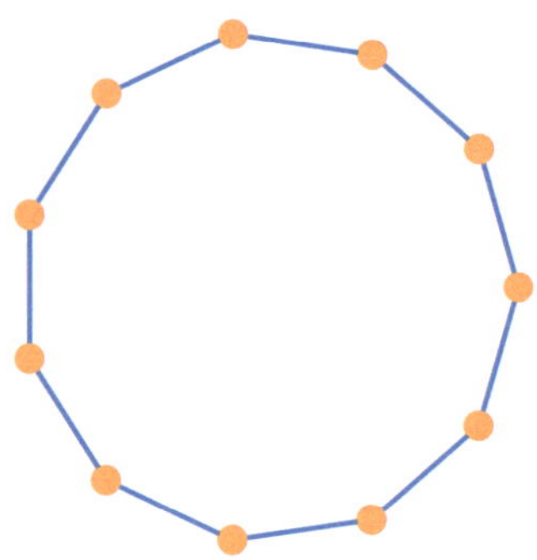

Figure 2.9: Ring lattice network

A ring lattice network has distance-based connections. As in a regular lattice network, the nodes are connected to all the neighbouring nodes which are in the range of a threshold distance d. The only difference here is the layout of the nodes, which are located in a circle.

- Random networks

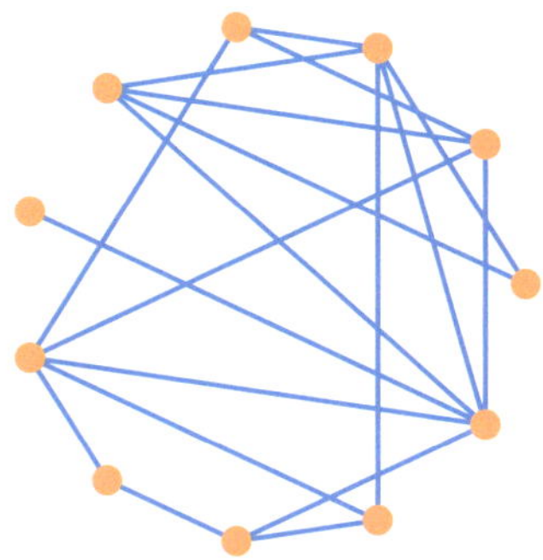

Figure 2.10: Random network

In a random network, the nodes are wired with certain probability p. Each edge, independent of every other edge, is included in the graph with probability p. For $p = 1$ therefore we would have a fully connected network, for $p = 0$ a network without connections. Random networks are located in between. Wiring edges randomly leads to a bell shaped Poisson distribution of the numbers of connections of each node, thus there are many nodes with a similar number of links. These networks were identified by Erdős and Rényi [1959]. The Erdős-Rényi Model was the first algorithm proposed to generate a random graph and we will refer to them as ER networks, too.

- Small world networks

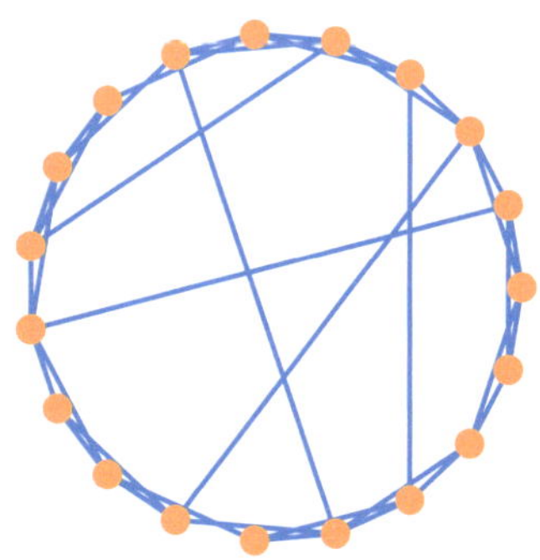

Figure 2.11: Watts-Strogatz small world network: the rewired edges can be seen as *shortcuts* which drastically reduce the average path length but still maintain a highly clustered network, which can be inferred from the originating ring lattice

In 1998, Watts and Strogatz found a new type of complex network. Having lattice networks which are distance based and show long path lengths, they began to rewire some edges, taking them out from the lattice network and replacing them with a random edge. If this is done enough times, it results in a random, ER-network. However, if only some edges are rewired, some shortcuts are created, which drastically reduce the path length among the nodes. The study of these kinds of networks leads to the commonly used term *small world*. A small world is a system in which some individuals within strongly clustered, but rather isolated, groups are connected with each other. This provides both highly clustered and also well connected systems. Any system with small average path lengths and high clustering coefficient can be called a small world.

The Watts-Strogatz (WS) model is an algorithm which generates a small world network. We start from a ring lattice network and begin to rewire random edges. As a random edge is selected, the end node is rejected and a new, randomly chosen node is selected. This rewired edge represents a shortcut to a usually remotely located part of the network part of the network. As the original edges are highly clustered due to their lattice character (they are distance based wired, so they form so called cliques, fully connected subgraphs), the connections among these clusters is not significantly affected by replacing the end node of one of the edges. On the other hand, the rewired edge connects over a short path the entire cluster to another, remotely located part of the network.

- Scale free networks

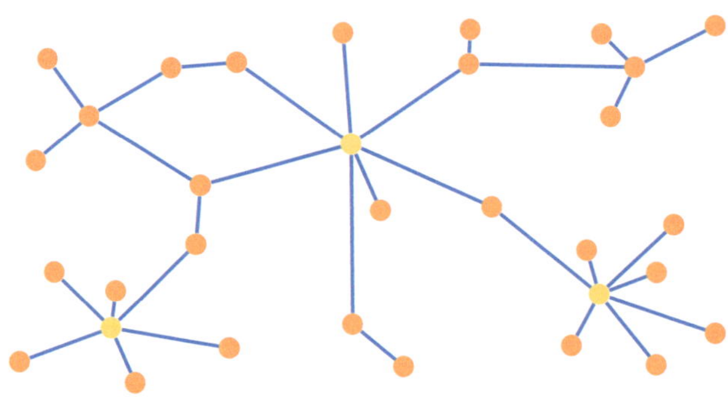

Figure 2.12: Scale-free network: Hubs are colored in yellow and have a large number of connections.

In real systems, for example in social networks, properties of random networks and small world networks were identified. However, at the end of the 1990's, Albert Laszlo Barabási and Reka Albert found that many real world networks could not be fully modelled by using these approaches. ER and WS networks have a Poisson degree distribution in common, which generally means that there are many nodes with the same number of connections, and few with large or small numbers of connections.

Barabási and Albert identified many systems with networks that have only a few nodes with a high number of connections, and many nodes with low number of connections. This is the case, for example, in the airport system: only some big airports offer a large number of destinations, whereas many local airports often serve only a low number of destinations. Another example is the western US electricity network, where a few big stations connect many lines, and there are a lot of stations with only a few connections.

This is reflected in the average network degree which is typically low, and in which the degree distribution looks different than in ER or WS networks. When *moving* a peak distribution (as appearing in ER or WS networks, see Figure 2.7) on the degree axis towards zero, we reach a system in which there are many nodes with a low degree of connection, and only few with many connections. While analysing these kinds of networks, instead of a bell shaped Poisson distribution, the degree of the nodes follows an exponential function. Barabási and Albert called these kind of networks scale free.

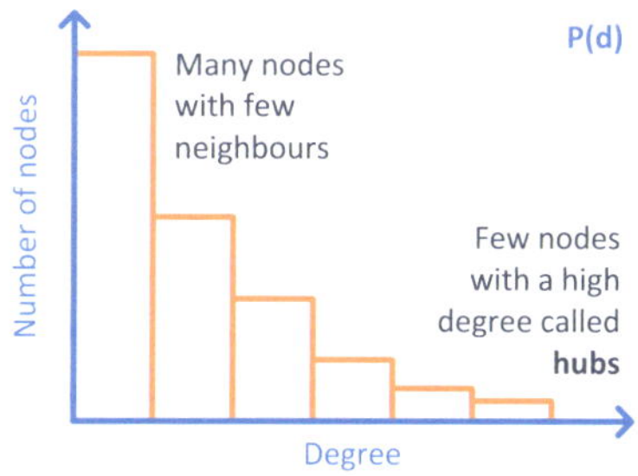

Figure 2.13: Degree distribution of a scale free network. The distribution follows a power law which leads to only few, very well connected *hubs*.

As it can be seen in Figure 2.13, there are many nodes with few connections, but only a few with a large number of links. These are so-called *hubs*. The degree distribution of scale-free networks follows an exponential decay and is proportional to the

$$P(d) \propto d^{-\gamma} \tag{2.8}$$

The Barabási-Albert model is a network algorithm which reproduces the scale free property by adding a node to a graph connecting it preferentially to nodes with a high degree (preferential attachment). The connection probability is proportional to the degree of the nodes. This represents the phenomenon *the rich get richer*. Barabási and Albert [1999] found that the internet, the airport system, or the power grid mostly have these properties

As we see, different types of network models have been found and are used to model the topology of complex systems. Figure 2.14 shows an overview and comparison of them.

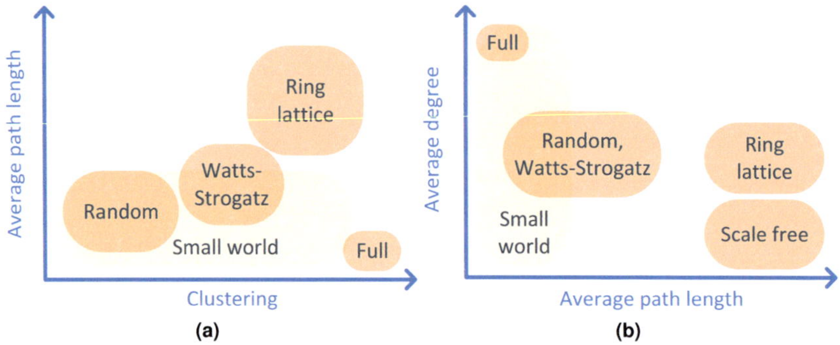

Figure 2.14: Overview of different characteristics of types of complex networks. In (a) the average path length is shown as a function of the clustering coefficient. In (b), the average degree is illustrated in function of the average path length. Both figures locate different types of complex networks based on their typical characteristics.

Complex networks are a formal approach to network science. The findings in this field (which is called statistical mechanics of complex networks by the physicists who study it) are rather theoretical and describe the properties of the networks in term s of their topology using graph theory. Many of these findings could be relevant to real and existing networks, and have been already studied through the social sciences or in relation to biological networks. In the area of energy, only few studies can be found. These focus mainly on the transmission system (by finding the scale-free property of a transmission network, for example), and on its robustness against attack, as we will see further in Section 2.2.4.

Where is the Complexity in Relation to these Networks?

One may say, complex network theory is mainly graph theory, which has already been studied by mathematicians and physicists. However, some important features, which relate these networks to complexity science, make the difference. Let us see the following example. As mentioned at the beginning of the chapter, complex systems are positioned somewhere between systems that are highly ordered and systems which are highly disordered.

This aspect of complexity can be shown through the Watts-Strogatz model. We start with an ordered ring lattice graph in which we rewire edges randomly. The average path length drops quickly after an initial limited rewiring, while the clustering coefficient remains almost constant, leading to a typical small world network. If we continue rewiring more and more edges, a random graph will emerge - a complete disordered system. Here we see that the average path length is still low, but has not decreased much since our *small world state*. It is noteworthy that the clustering coefficient dropped - another characteristic of a disordered system.

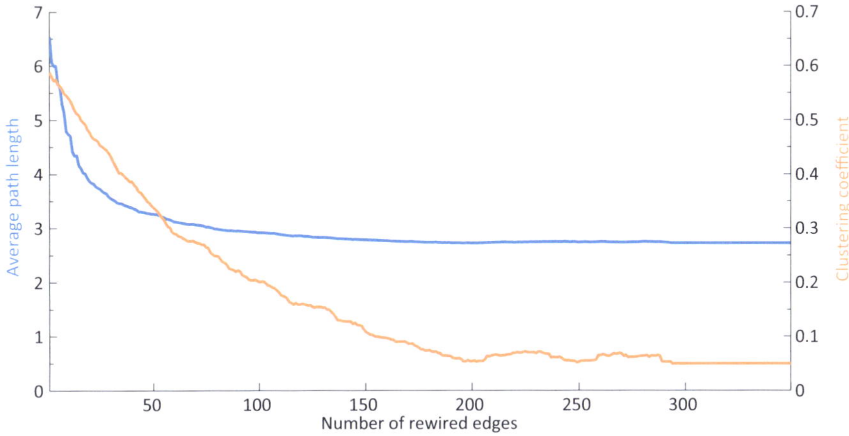

(a) Linear scale

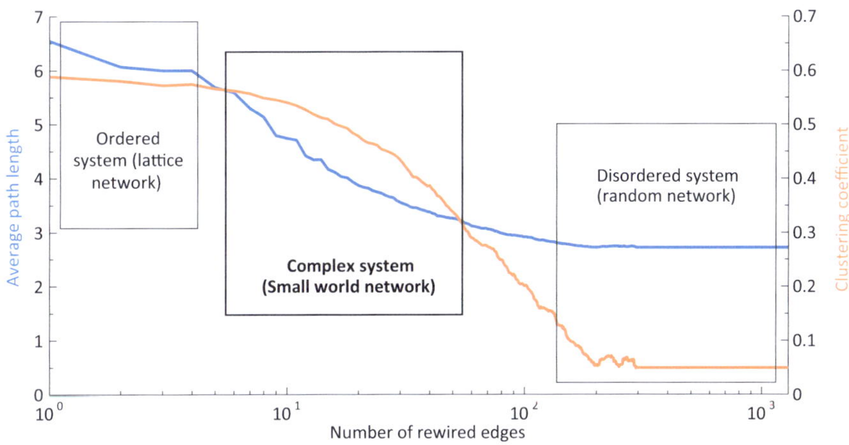

(b) Logarithmic scale

Figure 2.15: Complexity in ordered-disordered systems: Average path lengths L and clustering coefficients C for a Watts-Strogatz model with 99 nodes, where edges are rewired consecutively by reconnecting a *lattice*-edge to a random node. The x-axis represents the number of edges which have been rewired. Both values decay for increasing number of edges rewired, however, the range between a lattice graph (ordered system) and a random graph (disordered system) unifies special characteristics of the small world phenomena, short L and still large C at the same time. Complexity is positioned here, between highly ordered and highly disordered systems.

Complexity thus resides between an ordered network and a completely random topology. Here, the characteristics of both order and disorder meet and complex effects can be observed.

This example illustrates the location of complex systems and phenomena situated between ordered systems and randomness. The border of where a system gets chaotic is called the *edge of chaos*, a term introduced by Langton [1990] while analysing phase transitions on cellular automata. Langton introduced this term to refer to a critical point which separates order from disorder, and plays an important role in complex system theory.

Giant Components and Percolation in Network Theory

A giant component is a subgraph that is not connected to the other components and is disproportionately large when compared to them. When observing the creation of random graphs with the Erdős-Rényi model, first all nodes are isolated. After adding some edges, many small components appear. As we continue adding edges, at some time, larger components will form. After having added sufficient edges, we will end up with a single large component; no isolated components exist in the network.

A giant component typically emerges when surpassing a critical stage in the ER model. The giant component denotes a group of non-isolated, connected nodes which are relatively large compared with the other components on a graph. When many large components exist, at a given time they will become interconnected and form this giant component. This threshold is called the percolation threshold.

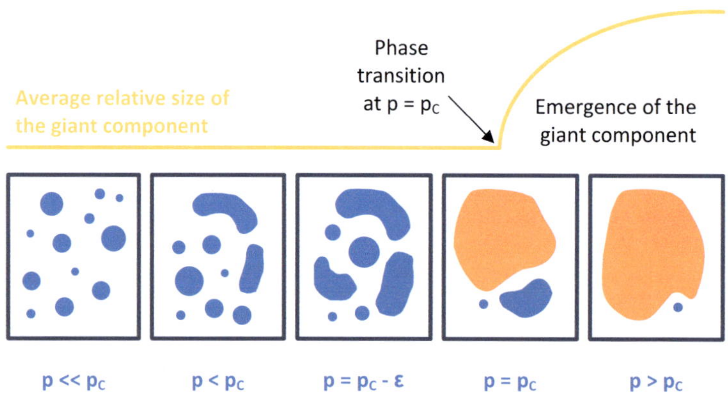

Figure 2.16: Emergence of a giant component. Randomly connected Erdős-Rényi network with probability p. With increasing p, the components of the network grow. While $p < p_c$ many separated components can be observed. Once p_c is reached, the giant component emerges. The system tends to change suddenly in a small range of p from a regime, where many independent, isolated components exist to a state in which a large number of nodes form part of a giant component. This is a typical percolation transition which happens at the critical threshold, where a phase change takes place.

Expressed as the random wiring probability p, a critical percolation threshold p_c, exists and, when reached, leads to the formation of the giant component. This emergent phenomenon is also known as a phase transition, due to the analogous principal observed in physical systems, that means a system that passes suddenly or in a short time from one regime to another (see Section 2.2.1).

An analog parallel can be traced with electrical or railway networks. Historically, when first introduced, those networks were small and isolated. Through time, they grew and expanded, and some of them were connected, for example to national grids. As they grew, a giant component emerged. An example of this is the European UCTE network, which interconnects a power system through a whole continent.

2.2.3 Power Laws

As can be seen, power laws seem to play an important role in complex systems. In the case of scale free networks, degree distribution follows a similar law. The percolation function is also related to a power law, as well as to city sizes, where there are only a few mega-cities with many millions of inhabitants, but a large number of smaller towns. If we draw the frequency of words in a novel or another lengthy piece of writing, we will see that some words appear very frequently and many others that are very rare. This is described by Zipf's law which is a special case of a power law distribution [Adamic, 2011].

Figure 2.17: Word frequencies in this thesis. The words are sized in proportion to their frequency of appearance in this document. Very common words such as "the" or "a" where removed. We can see that there are few very frequent words, but a large number of words which appear only once or very few times. The distribution follows a power law. [Created with `wordle.net`]

In Figure 2.17 the word frequencies of this thesis are shown. So, for instance, the words "system" or "model" appear over 300 times each, whereas words such as "experiment" or "computer" appear less than ten times. However, the amount of words that appear less frequently is much higher: there are over 5 000 words that appear less than ten times. In Figure 2.18 the word frequencies of this thesis are plotted in linear, half-log and log-log planes. The power law can be clearly observed.

In any case, power laws are being studied in many fields of science in order to know more about their origins and appearance in the real world. One of the main points of interest in power laws is their *scale invariability*: whereas other probability distributions change their

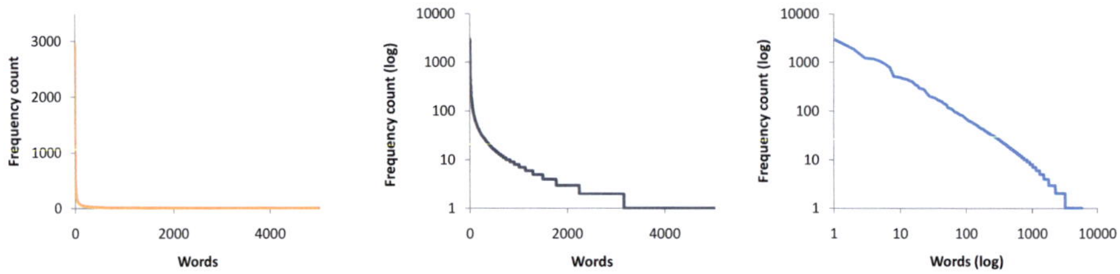

(a) lin-lin **(b)** log-lin **(c)** log-log

Figure 2.18: Power law shown by the word frequency in this thesis: in can be clearly observed that there are few words which appear very often, and many others that are less frequent. In the linear plane (a), on the right there is a long tail, and on the left, few words with a very high count. In the log-log plane (c), the function has almost a linear shape.

shape over scale, a power law distributions shows a perfect straight line on a log-log plane. This means that throughout all the scales of the system, the same structure and relations can be found – this is also known as *fractal scaling*. This scale invariability is quite important, as it allows us to find relationships in multi-scale systems which can describe the system as a whole, over many orders of magnitude.

2.2.4 Resilience

Resilience describes the robustness of a system against disturbances. Systems need to compensate or withstand disturbances coming from the interior or exterior while maintaining the integrity of the system. In network theory, resilience can be shown by analysing the effects of node removal on the system. This is closely related to the phenomenon of percolation (Section 2.2.1), but instead of adding edges and observing the emergence of a giant component, here we remove edges which corresponds to random failure of links. The percolation threshold, also called critical probability, appears when the removal of links leads to a breaking of the giant component. The threshold is reached when the average degree K of the nodes is below 1 and the network decomposes into several components.

Albert, Jeong, and Barabási [2000] analysed the behaviour of networks concerning node removal and discovered that scale-free structures are resilient to random failures, but sensitive to targeted attacks. For random networks, there is less difference between the two types of attacks. Lattice or highly ordered networks are quite vulnerable to random failures. This is especially interesting in networks like the power grid, which are likely to be scale-free. Other studies confirmed these theoretical predictions on the North American power grid [Chassin and Posse, 2005]. Rosas-Casals et al. [2007] and Solé et al. [2008] analysed the robustness of European UCTE power grids under a targeted attack, comparing it to random node attacks. They found that the UCTE network does not mostly fulfil small-world properties, as there are long path lengths from one country to another over the

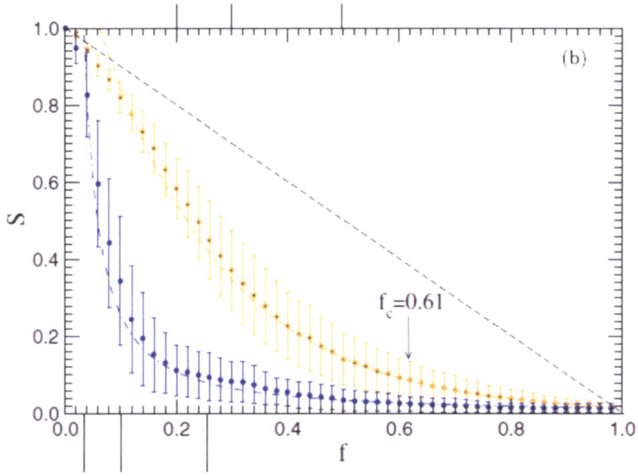

Figure 2.19: Vulnerability of the power grid to random and selective attacks, based on the relative size S of the giant component in function of a fraction of nodes removed p. The orange plot shows the removal of random nodes of a part of the UCTE network, whereas the blue one represents a selective removal of the most connected nodes. It can be observed that the giant component size decays much faster for a selective removal i.e. the network is more vulnerable to a targeted attack. Source: UK and Ireland network analysis in [Rosas-Casals et al., 2007]

whole continent. The response of the network is resilient to random failures, whereas the same cannot be said for selective attacks, where nodes with high degree are targeted (hubs or strategically important nodes).

As discussed, resilience depends strongly on the topology of a network, as well as on the type of disturbance (random or selective) considered.

2.2.5 Controllability

Complex systems can be characterized by many phenomena that seem to be common to them, even though the nature of the system might be completely different. The existence of complex network topologies within those systems, as well as phenomena such as emergence or system properties as resilience, have already been explored. These studies deal with an analytical and explorative point of view of the system. This approach is mainly inherited from natural sciences where the complex system theory allows for the better understand of their behaviour. Biological systems, such as a cell which can be seen within a compound or aggregate and forming different modules like organs within an organism, is truly a complex system. These systems can be explored and can probably be better explained due to advances in complex systems theory over the last few years.

However, in an effort to apply these findings to man-made systems, the questions are rather different. Exploration might be a topic, but as the system itself was engineered by man, its fundamental mode of operation and its structure is well known. This is not the case in biology. In addition, man-made systems are conceived to pursuit an aim. Therefore, it seems important to ask how the system works, and also how we can influence its operation in order to better fulfil the goals for which the system was created.

Recent studies have shown that there seems to be a common set of findings related to *controllability* throughout a diversity of complex systems [Liu et al., 2011]. Control theory defines that a system is controllable if it is possible to drive it from any initial state to any desired state within a finite time. In the case of a complex network, the time dimension has to be introduced to allow the description of the state variation of the nodes. This means that the network evolves over time and that this involves a change in state for the nodes. In order to control the system, a signal is applied to a specific node. To control the whole system one must be able to influence all of the nodes. If we can identify the set nodes that fulfil this property, we will be able to take control of the system by only acting on these nodes. Liu et al. have called them *driver nodes*. They analysed different properties of these nodes and found, for example, that the amount of driver nodes can vary significantly depending on the type of network. Social networks seem to have few of them, suggesting that only a few individuals could control the system, whereas in gene regulatory networks a high number of nodes (around 80%) are driver nodes, and thus needed to be controlled in order to manage the whole system.

2.3 Electrical Energy Systems

The electrical energy system is composed of different electrical components that allow for the production, transmission and consumption of electric power. Production or generation of electrical power is the process in which other energy sources are converted into electrical energy. For transmission, different kinds of electrical networks are used. These networks are generally classified by their nominal voltage at each level. Long range transmission over hundreds of kilometres is performed at high voltages of several hundreds of kilovolts (kV). These networks are called transmission systems. Transformer stations (substations), can reduce the voltage at a given point in order to feed electricity to the so-called distribution system. The distribution system carries the electricity to the consumer. These networks operate at medium voltage levels, usually between 1-50 kV. In a final stage, distribution transformers can convert from medium voltage to low voltage (less than 1 kV) which is the typical voltage level found at residential or tertiary customers. Some specific customers (such as industries) may have direct connections at the medium voltage level.

In a *classical* energy system, generation is injected at high or medium voltage levels and consumed in the distribution system. Due to the introduction of renewable energy sources and distributed generation, energy can be injected at almost any level of the system. Also, and in order to better match the fluctuating production introduced to the system, Demand

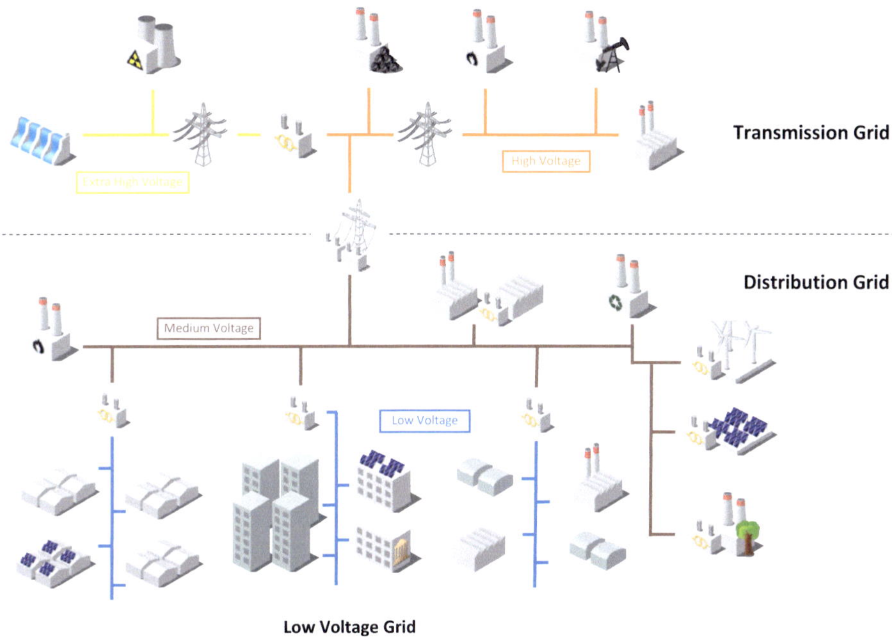

Figure 2.20: Electrical energy system: the different parts of an electrical energy system are shown, which is constituted of the transmission grid, the distribution grid and the low voltage grid. The different voltage levels are indicated for each of the networks.

Side Management (DSM) mechanisms are being developed. These DSMs allow the management of consumption and ensure a more sustainable operation of the whole system.

The electricity system can be seen as a complex system, composed of a large number of interacting entities. Reproducing the behaviour of the system is therefore not possible by only modelling individual objects or by modelling the system in a monolithic way. To understand such complex systems, and to improve their design, it is not sufficient to study their components separately, using the specific formalisms these components were modelled in. It is necessary to answer questions about properties, most notably behaviour, of the system as a whole [Vangheluwe, 2008].

2.3.1 Electricity Generation

Electricity generation is the process of converting energy in other forms (chemical, mechanical, nuclear, etc.) into electrical energy. It is the first process for electrical utilities in the chain of the delivery to the consumer. Electrical production can be performed in many ways, depending on the source of energy used, as well as the size of the production unit.

Two mayor classifications of electricity generation, must be considered when looking at the current state of the energy system.

The first characterisation is related to their size and hierarchical grade in the system. Centralised power plants tend to be located in remote areas with suitable conditions for their deployment (riverside for cooling, well connected to primary resource supply, etc.). This is the traditional way of production that has been the foundation of the system since electrification. Classical production relies on large production units, feeding into a transmission grid to transport energy over long distances and distribute it to the final consumer, in a tree-like structure. The leaves would be the final consumers, which are fed from a central source in the roots. In this configuration, flows are almost only uni-directional, from the source to the sinks of energy – from the roots to the leaves.

The first distributed systems date back to the 1950s, when local generation was first considered as an alternative to centralised structures. Distributed Generation (DG) avoids transmission costs and losses, by producing the energy close to its point of consumption. However, this isn't possible with all types of energy sources. In the classical paradigm, the location of coal plants outside an urban area was to avoid local pollution. Renewable energy sources first began to be integrated as they are environmental-friendly and can be deployed as small DER. Table 2.1 shows an overview of the different types of generation classified by their degree of hierarchy.

Centralised	Distributed
• Power stations managed by utility • Large installed power capacity • Few production plants • Can be far away from consumer • Classical production follows this structure	• Energy production based on inter-connected little and medium size power generators and/or renewable energy plants • Small or medium installed power capacities • Large number of production units • Local production for local consumers

Table 2.1: Centralised vs. distributed generation

With a distributed generation, electric energy is produced close to the consumer, within or near residential areas and industrial plants, through small production units. The dimensioning of the power generation equipment is usually only designed to meet the energy needs of locally connected consumers. Excess energy can be fed into the distribution or transmission grid. The transition to a smart grid will first involve the introduction and then the expansion of distributed generation which will result in massive use.

The second aspect which allows us to characterise generation is the primary energy source type. Here we distinguish between *conventional* and *renewable* energy sources. Conventional sources have been used since the beginning of electrical energy generation. They usually include fossil fuels and, more recently nuclear fuels, to power thermal plants. The process is based on steam generation moving a turbine converting energy from mechanical into electrical form. Renewable energy sources have gained more and more importance during the past decades mainly because of environmental reasons and because of finite fossil fuel reserves. They include thermal generation through sustainable fuels, which operate similarly to conventional plants, and other generation methods, such as wind power where wind turbines directly convert wind into electrical energy. Other means of generation use diverse methods, such as solar photovoltaic panels through the photovoltaic effect converting radiation into electricity, or the use of tidal energy to generate electrical power. Table 2.2 shows a synthesis of the generation types by primary energy source.

Conventional	Renewable
• Thermal production based on fossil fuels (usually coal, petrol, gas, etc.) • Thermal production based on nuclear fuels	• Thermal production from sustainable fuels (biomass, biofuel, biogas, etc.) • Wind power • Hydropower • Solar power • Geothermal power

Table 2.2: Generation by primary energy source: examples.

Renewable energy sources (RES) must not be confused with distributed generation. Renewable energy is usually associated with distributed plants. However more and more large, centralised renewable production plants are being built (large wind farms, solar farms, etc.). On the other hand, there are also conventional production systems which are distributed (e.g. micro CHPs).

The introduction and, particularly, the expansion of renewable energy sources leads to important questions. Is the necessity of covering the demand assured, as RES tend to fluctuate (e.g. wind) or may only be periodically available (e.g. solar, only during daylight)?

Again a trend towards the smart grid can be recognised in this classification. Generally, the introduction and achievement of larger penetration rates of renewable production units can be observed. For example, in Spain in the year 2009 already half of electrical consumption at a specific moment was covered by wind power energy [Red Eléctrica Española, 2009].

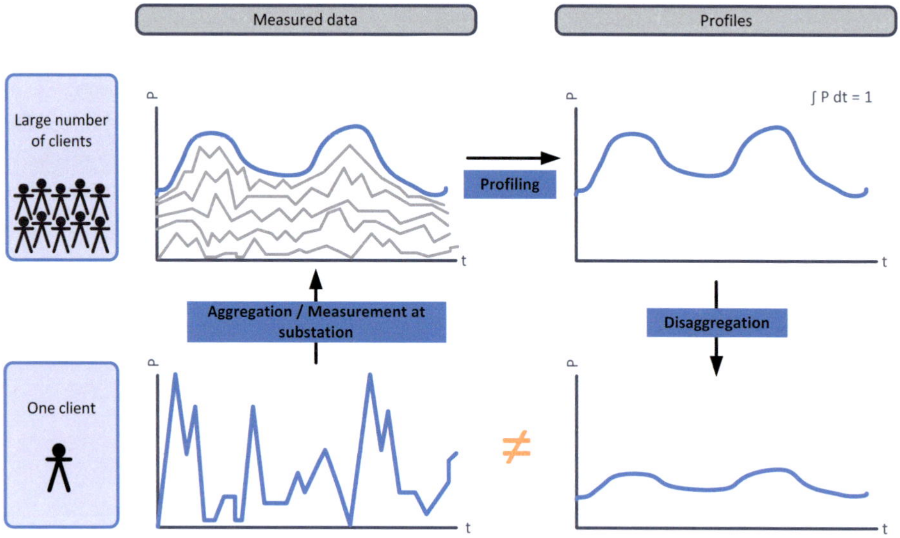

Figure 2.21: Standard load profile method limits: why profiles don't work for individual customer load curve estimation.

2.3.2 Electrical Energy Consumption

Electrical energy consumption can be split into productivity sectors, such as industry, transportation, commercial and residential use, etc. Industry consumes the greatest amount of energy worldwide, followed by the residential and commercial sectors. Consumption is usually not well planned. Electrical energy consumption over a period of time is characterised by its load curve. A load curve describes the $P(t)$ function, which is the momentary power consumed at a given time t. When integrating the load curve, we obtain the energy consumption over a period of time T, $E = \int_{t=0}^{T} P(\tau)d\tau$.

Load curves for a large number of consumers, usually show typical shapes, characterised by load-use. So, for example a typical load curve of a residential area shows morning and evening peaks, which coincide with the increased use of electricity for cooking or lightning, for example (Figure 2.21).

By estimating upcoming electricity demand, it is possible to plan production resources. Forecasting consumption allows an optimised generation of electricity, for example by using base-load power plants which are cheaper and by using highly reactive, but more expensive, generation units only at peak times. Production can be better planned better, when demand is better known.

Prediction of Electrical Consumption

The standard load profile method is commonly used for aggregated prediction. It relies on load profiles measured, or estimated, for specific consumers. These can be scaled for modelling demand. This approach was chosen for industry/business and the public sector, as most of the available data was aggregated and the behaviour of these sectors is based on similar patterns. Synthetic load profiles are the result of a statistical analysis based on representative samples from different consumer groups [Palensky et al., 2008].

A standard load profile is a representative load profile that is used to predict the demand of a user without making use of measurement. The profile is a normalised curve for a specific consumer, consumer group or appliance that was obtained from measurements or estimations of many of these consumers. It represents the demand behaviour of a large and aggregated group of users that is to be considered. This is especially helpful for creating universal profiles for consumers that have heterogeneous behaviour and cannot be described by aggregating the same individual entity many times over. This represents the average demand behaviour of this consumer group. By the normalisation of the curve, it can be scaled further and customised for the model in which it is applied.

Load profiles are normally provided in normalised form, and the user has to multiply them by the total amount of energy resulting from the profile. This is normally daily, or yearly consumption, for example. As energy amounts for electrical use usually result from surveys and studies, having the profile is a practical way to obtain an approximate load curve. The approach is applicable in a broad range of uses and does not involve too many complex models. This makes it quick and results in good estimates at aggregated levels. Nevertheless, it has to be kept in mind that these curves can only reflect aggregated behaviour and are not representative if individual users or small groups of users are considered [Willis and Scott, 2000; Paatero, 2009].

A further issue of current demand prediction or simulation models is that they mostly rely on low resolution data (e.g hourly sample rates). Hourly averages are suitable for modelling the demand at a high and aggregate level. However, when dealing with individual consumers, there is a need for high resolution data [Wright and Firth, 2007]. Only a few high resolution studies and models exist on domestic demand Wright and Firth [2007]; Widen et al. [2009], and they are not coupled to an energy system model.

These curves are not suitable for planning and control at the distribution level. They only apply to a large number of consumers as they describe the average use of energy over time [Willis and Scott, 2000; Paatero, 2009]. Also, when considering distributed generation or renewable energy sources, production cannot be accurately represented by this method. Diversified profiles are only applicable to community generation with a single point of import and export upstream of a large number of houses and the generator - but standard profiles are unlikely to apply to a particular community [Wright and Firth, 2007].

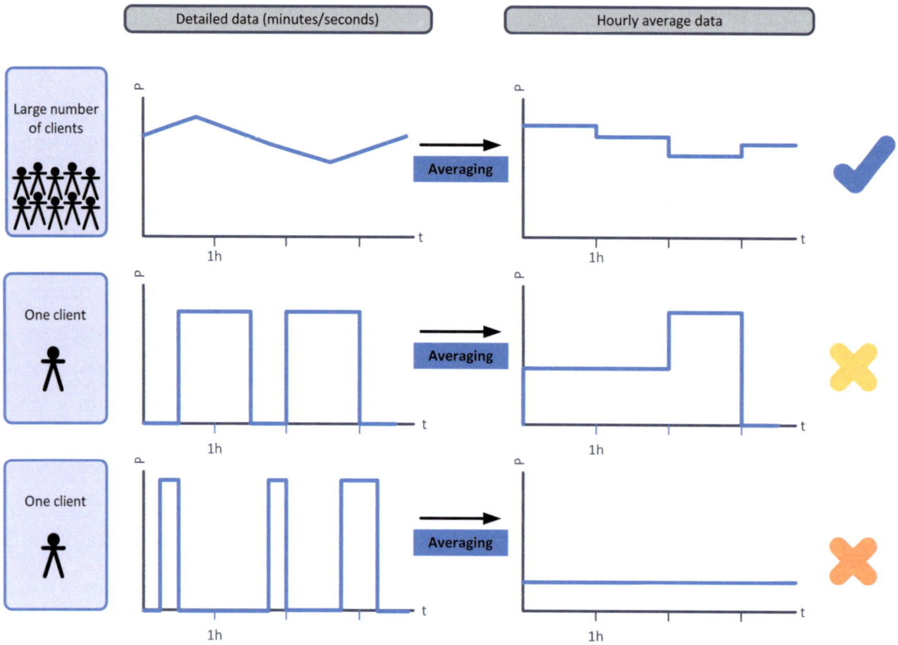

Figure 2.22: The average effect: why high resolution data is needed if an individual approach is chosen. For large numbers of consumers, the load curves do not fluctuate too much. For sharp peaks however, as it is the case for individual consumers, average values distort the load curve significantly.

Metrics used for Load Curves

In order to quantify a load curve, which is the load consumed over a time period, the following metrics are used:

The load factor The shape of a load curve or profile can be analysed with the load factor, which is the ratio of the maximum power to the mean power [Jenkins et al., 2000]:

$$LF = \frac{\frac{1}{N}\sum_{t=0}^{N} P(t)}{\max_t P(t)} \tag{2.9}$$

The variation of the load factor is in the range between zero and one. High values are an indicator for a smooth curve without large variations. A low value, on the other hand, indicates that there are high power variations. The peak value is high in comparison to the mean value of the curve in that case.

The coincidence factor When individual loads are added, they can coincide to different degrees. If many of the loads have their peak at the same time, this will be reflected in the aggregated load curve. If the peaks are distributed and do not coincide, the load curve is smoother. This effect can be described and quantified by the coincidence factor [Jenkins et al., 2000]:

$$CF = \frac{\max_t P(t)}{\sum_i \max_k P_i(k)} \tag{2.10}$$

where an aggregated curve $P(t)$ is contemplated, which is formed by $P(t) = \sum_i P_i(t)$ where P_i are the individual loads. If the individual maxima of the loads occur at the same time, the coincidence factor equals one, thus having a high coincidence or synchronisation. If the peaks of the individual loads are distributed over time, the coincidence factor is lower, indicating a smoother curve.

The coincidence factor is a measure over a load curve. To define the coincidence of loads at a given time t we define the coincidence

$$c(t) = \frac{P(t)}{P_{inst}} \tag{2.11}$$

where $P(t)$ is the instantaneous load and P_{inst} the total installed load (maximum possible load if all individual loads would coincide).

Demand Side Management (DSM)

In the future, smart electrical system, the demand side will have an active role. By regulating demand within certain limits, not only is production governed but it also allows for a more flexible balancing of the system. The use of demand side management (DSM) has been widely discussed for different purposes in enhancing performance and avoid critical situations for the power grid Palensky and Dietrich [2011]; Albadi and El-Saadany [2007]. Management of the demand side would allow for:

1. Improving energy efficiency
2. Reducing peak loads
3. Shifting loads over time
4. Increasing valley loads
5. Stabilising grid frequency

The first possibility represents a lasting improvement to the demand side (for example, permanently reducing demand through increased energy efficiency, by, for instance, using energy-saving lights). Points 2-4 focus more on load curve shaving, acting in a medium term (daily or at most hourly) view, which is also called demand response (DR). These mechanisms can support the integration of fluctuating renewable energy sources such as wind power or solar generation. The last possibility allows the support of the primary or spinning reserve. By acting to support the primary reserve and therefore assure stable frequency levels in the grid, the actions occur within a time windows of some tenths of seconds. This reserve is the first one that acts on a grid disturbance, such as a sudden generator unit failure. In this case, it is important to react rapidly by injecting additional power into the grid. This avoids a critical drop-down in frequency and temporarily restores the production-demand balance. These measures are taken automatically within the first seconds after the grid disturbance, by the production side. Most of the generators are fitted with automatic speed control. Once the frequency is stabilised, manually scheduled increases in power (secondary reserve) or dispatching additional units are performed in order to replace the rest of the lacking load.

2.3.3 Modern Energy Systems

Smart Grids

According to the European Technology Platform on Smart Grids, a smart grid is an electricity network that can intelligently integrate the actions of all users connected to it – generators, consumers, and those that do both – in order to efficiently deliver sustainable, economic and secure electricity supplies.

The term smart grid has been used since 2005 [Amin and Wollenberg, 2005], although with somewhat different meanings, but all of which give scope to two key elements: digital data processing and communication networks. Therefore it can be said that what characterises the intelligent grid is the existence of a flow of data and information, between the supplier company and the consumer, running in parallel with the flow of energy [Singer, 2009]. This concept is expanded by some authors up to a complete inclusion of (micro-)SCADA systems at all levels of the grid. These industrial control systems (SCADA: Supervisory Control and Data Acquisition) allow the monitoring and controlling of industrial and infrastructure-based processes, such as the electricity system. They have been widely implemented on the transmission grid for decades.

The term smart grid usually covers the whole spectrum of the electrical system, reaching from transmission over distribution up to the final customer. As such a broad area is implied by this term, there are many sub-fields of research that are related to the topic, but do not only involve the voltage level of the grid. This is why the smart grid is used to refer to the *smartening* process of the electricity system as a whole, while focusing on all of the parts on the system that have to be capable of communicating, using information technologies (IT) or in general, have some kind of intelligence (although the definition of this term is not yet agreed and very vague).

The smart grid is intended to improve the provision of electricity from suppliers to consumers throughout the deployment of information and communication technologies (ICT). An internet like network establishing communication as well as special digital electronics to control energy generation and consumer demand are needed to set up this true complex system. It will allow suppliers to remotely monitor client power consumption, implement variable energy costs, and in this way avoid long peak load periods or circuit overload situations. Also, consumers will be able to monitor their real time power consumption, which could be used to load control by controlling their own peak and off-peak use in order to benefit from lower tariffs.

With today's smart grid goals in mind, energy supply companies are in a transition between the existing electricity grid and the future smart grid. They seek to improve the conventional network infrastructure, and establish a digital level (essence of the intelligent network) and also create new business processes to carry out the capitalisation and commercialisation of the intelligent network.

Generally, it can be said that the introduction of the smart grid involves an introduction of SCADA at almost all levels of the grid. SCADA systems have existed for years in the transport network. In order to allow effective system management at all levels, including technologies like SCADA into the distribution grid is necessary.

Other relevant technologies and mechanisms which are commonly mentioned in relation to smart grids are:

- Smart metering
- Electric vehicles
- Distributed generation
- Demand side management
- Energy storage
- Dynamic pricing

All these could be part of a future power grid, but non of them alone should be considered as an integral smart grid. The introduction of ICT will allow the management and control of most of these technologies. These technologies can also support the ICT infrastructure, for instance smart meters.

Smart Grid Demonstration Projects

Several demonstration projects are already running or are in preparation to show the possibilities of smart grids. A large catalogue of more than 200 smart grid related projects in the EU is presented in [Giordano et al., 2011]. The distribution of projects across Europe is not uniform. Most of the projects and investments are located in *old* EU-15 countries, while the newer EU-12 member states still lag behind. These projects have some common challenges which must be tackled. A multidisciplinary cooperation is needed. Consortia involving multiple disciplines will contribute by sharing competences and reducing risks on projects dealing with the increasing complexity of the electricity system.

There is a trend towards the fruitful cooperation of different organisations, bringing together network operators, academia, research centres, manufacturers and IT companies. The implementation of smart grids will also be significant opportunity for European industry to research, market and export new technologies and to maintain global technological leadership.

Most smart grid benefits are systemic in nature as they arise from the combination of technological, regulatory, economic and behavioural changes. Giordano et al. [2011] indicate that in almost all countries, a significant amount of investment has been devoted to projects addressing the integration of different smart grid technologies and applications. The concerned technologies are mostly known, but the challenge lies in integrating these technologies.

ICT are crucial for the development of smart grids. An open and secure infrastructure for ICT is one of the principal requirements for the success of smart grids. Standardisation to ensure interoperability and further research on data acquisition and analysis are of equally high priority.

The Two Layer Model

The smart grid can be considered as a system capable of handling and managing a true complex electricity system. When the system is abstracted, it consists of a set of vertices (nodes or buses) with two types of edges arranged between them: the first one (power lines) is dedicated to carrying the power flow and the second one (information channels) for broadcasting the information flow. Therefore, the electrical system can be divided into two layers:

- Physical layer: The first layer is the physical structure of the electrical grid itself, including all the power equipment such as transformers or transmission lines. It comprises the power flows as well as all the electrical parameters related to the correct operation of the grid.
- Logical layer: This second layer, which represents the main part of the future generation of electrical grids, is not yet completely implemented into all levels of the system, in contrast to the physical layer. It can be seen as a complementary layer, built around the first one. It hosts the information exchange that has to be arranged to control distributed resources, dispatchable loads and all other *smart equipment* in the future grid. It must be emphasised that the communication paths do not have to be the same as the links in the first layer, although they could be exploited for that aim. An example of this is Power Line Communication (PLC).

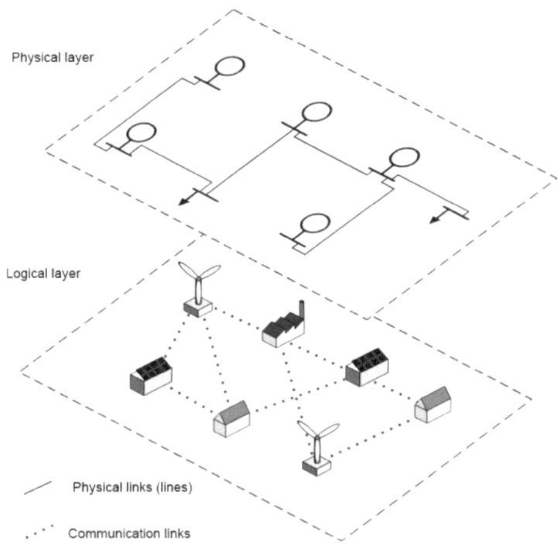

Figure 2.23: The two layer model of a smart grid

There are several implementation possibilities for the second layer due to the wide spectrum of communication technologies available. It is assumed that this layer will implement a system that allows real-time communication between the elements of the grid. In the German E-Energie project[2], an *Internet of Energy* is suggested as an analogy to computer networks. This medium could itself serve as a communication platform. More examples of the network implementations for the logical layer are:

- Wired technologies
  - Using the power grid infrastructure: PLC
  - Internet infrastructure: over existing networks such as telephone or cable networks
  - Wired solutions on a (new) dedicated network
- Wireless technologies
  - Internet related technologies: WLAN, WMAN or WiMax
  - Mobile technologies: GSM/GPRS or third generation networks such as 3G/UMTS
  - Lower range technologies: Bluetooth or Zigbee for local communication

The communication layer also supports other smart grid relevant topics such as geographic network localisation, distribution of processing and databases, interaction with humans, and unpredictability of system reactions to unexpected external events, among others.

[2] `http://www.e-energy.de/`

Microgrids

A microgrid is an integrated energy system consisting of interconnected loads and distributed energy resources which, as an integrated system, can operate in parallel with the grid or in an intentional island mode [Navigant Consulting, 2006]. A microgrid is also a set of small energy generators arranged in order to supply energy for a community of users in close proximity. It is a combination of generation sources, loads and energy storage, interfaced through fast-acting power electronics. Emerging from the general trend of the introduction of renewable energy sources (RES), microgrids will mostly include this type of generation, so they form part of the hybrid renewable energy systems (HRES). These systems combine two or more energy conversion devices, or two or more fuels for the same device, that when integrated, overcome limitations inherent in either. Microgrids represent a form of decentralisation of electrical networks. They comprise low- or medium-voltage distribution systems with distributed energy sources, storage devices and controllable loads.

The European Commission [2011] defines microgrids as small electrical distribution systems that connect multiple customers to multiple distributed sources of generation and storage. It also states that microgrids can typically provide power to communities of up to 500 households at low voltage level. This shows that this technology is not only suitable for the electrification of small remote islands or remote zones in developing countries, but also for integration into existing high voltage transmission grids. Therefore microgrids could have a wide application spectrum in the near future.

During disturbances, generation and corresponding loads can autonomously disconnect from the distribution system. This allows for the isolation of the microgrid load from the disturbance without damaging the integrity of the transmission grid. This mode is called the *islanding* mode. From the customer's perspective, it can be seen as a low voltage distribution service with additional features like an increase in local reliability, the improvement of voltage and power quality, the reduction of emissions, a decrease in the cost of energy supply, etc. A microgrid is connected to the distribution network through a single Point of Common Coupling (PCC) and appears as a single unit in the power transmission network. Power electronics will be a crucial feature for microgrids since most of the power micro-sources must be electronically controlled to gain the characteristics required of the system. The microgrid is therefore not only a more or less autonomous part of the power system, but also has to be a smart system itself. It has to be able to cope with multiple issues, as described below.

The introduction of smart grids involves a change from manual operations towards an intelligent, ICT-based and controlled network. These changes will especially affect the distribution grid [Jiyuan and Borlase, 2009]. Microgrids can support this change, by being a flexible and autonomous module of the smart grid. The challenges of implementing microgrids mean that they are implicitly considered to be smart grids, or at least to be part of a smart grid.

The operation of a smart electrical system will involve a higher degree of complexity than the operation of the conventional power grid. In order to function, components like automation, sensors, remote controlled switching devices and communication networks are necessary. For example, the current power grid is still not ready to permit microgrid connections. Connections made at present are experimental and almost always performed manually, by ensuring that a number of factors are fulfilled, for instance before completing a connection.

Microgrid Architectures In terms of their architecture, microgrids can be classified in four different types [Navigant Consulting, 2006]:

- Single facility microgrids

  These microgrids include installations such as industrial and commercial buildings, residential buildings and hospitals, with loads typically under 2 MW.

- Multiple facility microgrids

  This category includes microgrids spanning multiple buildings or structures, with loads typically ranging between 2 and 5 MW. Examples include campuses (medical, academic, municipal, etc), military bases, industrial and commercial complexes.

- Feeder microgrids

  The feeder microgrid manages the generation and/or load of all entities within a distribution feeder, which can encompass 5-10 MW. These microgrids may incorporate smaller microgrids – single or multiple facility microgrids- – within them.

- Substation microgrids

  The substation microgrid manages the generation and/or load of all entities connected to a distribution substation, which can encompass 5-10+ MW.

Island Systems

Interconnected continental power grids such as the UCTE or the North-American power grid offer a large degree of resilience due to their extension and number of control mechanisms. Further, generators and loads are geographically and topologically distributed over thousands of kilometres. The frequency of these grids is usually largely stable and only small variations occur during regular operation. For example, the effect of a failure in a production unit is low when compared to the overall production of the whole system. Thus, its effect on the frequency can be almost neglected. In addition to that, the usually sufficient reserve units are available to rapidly compensate for the dropped power station.

On island systems however, the resilience is drastically reduced. Small, self-sufficient electrical systems have been traditionally supplied with highly reactive, rather small or medium sized production units, such as gas or diesel plants (combined cycle), or even diesel motors. The smaller the system, the more difficult it becomes to control, as the shaving and smoothing effects of large systems are not present. Usually, a failure of one plant supposes an already large reduction relative to the total power consumed by the system. On the other hand, connecting large consumers, such as an industrial plant can cause similar imbalances.

Following the general trend of introducing renewable energy sources, fluctuating producers are being installed on island systems [Tarkowski and Uliasz-Misiak, 2003]. These systems are usually suitable, as their climate condition allows significant exploitation, for example due to their wind characteristics. The inclusion of fluctuating energy sources adds an additional variable to the system, which impacts directly on its stability.

Modern island systems are currently being used as a kind of *experimental labs*. They have specific issues, but they also offer an isolated, autarkic grid, which can be used as a test bench for new technologies. Some of these systems offer the opportunity to test large penetration rates of RES. The first island totally supplied by RES will implemented in El Hierro, Spain, where wind power coupled with a pumped hydro storage plant will allow independent and completely autonomous generation.

2.3.4 Storage Systems

The implementation of storage systems in energy networks offers new paths to explore. Storage systems offer flexibility in a system that has previously been mainly characterized by unidirectional units, which were, on the one hand generation units and on the other consumption units. In a power system, generation must always cover demand. Furthermore, not only must demand be covered, but no additional energy should be produced to avoid overcharging. A balance between production and demand must be assured at all times to guarantee a stable system.

Generation sources in conventional power systems are dispatchable (they can be controlled to gain a certain level of power output). They use different types of plants, like base, middle and peak load plants, in order to meet current demand. For new emerging systems, which include several types of fluctuating renewable energy sources, as well as distributed generation, this task becomes more complicated. New measures are needed to assure system stability. Intermittent energy sources cannot be or can only be partially, forecasted. Sufficient reserve power must be available at conventional or dispatchable plants to maintain production in case of abrupt drop-downs in the intermittent sources. But also, there might be overproduction periods, where too much renewable energy is generated without having a consumer to use it when available. Therefore, storage systems have gained in importance. They can be used to react to short term, transient effects caused by intermittent

(a) Classical electricity system

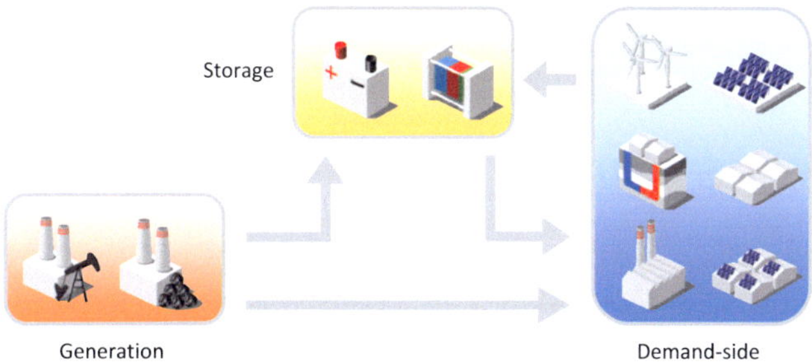

(b) Modern electricity system

Figure 2.24: Storage inclusion in modern electrical systems: In (a), a classical electricity system is shown, with unidirectional flows from generation to demand. Below (b), a modern system, in which the demand side becomes much more flexible by including production and demand side management. In this case, the energy flows can be in different directions. Storage can help to temporarily store energy between generation and demand-side.

sources and thus help to stabilise the system. Further, they are able to accumulate unused energy and deliver it at a later point, to where it is needed. Here, the strengths of these systems become really clear.

Moreover, the system's electrical demand is not constant through time. This generates periods of higher demand creating a need for increased generation, or periods of lower demand times where some units have to be dispatched at a low level or disconnected from the grid. To optimise the use of big generation units, it is important to function at or around the optimal point of operation. At this point it is possible to reach an optimal value considering costs, emissions, etc. But fluctuating demand will force at least one plant to operate at a different power level. Solutions to this are the so-called smoothing of the load curve, a technique to guarantee more constant production. More constant production curves usually reduce production costs, as the additional, expansive peak power plants are not frequently used. To achieve a flatter load curve, storage systems can be used, to store energy during low demand periods and deliver it at periods of high demand.

Storage systems are hybrid units that can function as production or consumption units, depending on network needs at any given time. Adding their features to a power system can make this system more flexible and improve resilience. This makes the system more stable when facing unpredictable changes and exterior influences of any kind. Electrical energy can only be stored in electrical form with difficulty. This is only possible in capacitors or supra-conducting coils, also called superconducting magnetic energy storage (SMES). Usually it is more economical to convert it into other forms of energy and then convert it back when needed. Each conversion involves losses, and the energy is lost over time during storage, too. The sum of all individual losses may be significant and make the process uneconomical. In Appendix B an overview of different storage technologies as well as typical values for their performance, capacity and other factors, is shown.

2.4 Energy Systems: a Complex Systems Approach?

Complexity science is a new field of science, which aims to understand our complex world from a new perspective. It can describe phenomena and effects that are common in many real systems, which at a first glance seem to be completely isolated from each other (emergent phenomena, for instance). A complex system approach allows the detection of common patterns, such as networks or interactions, which, independent of the system, will show recurrent properties. Studying phenomena at this level permits understanding our world from a more systemic perspective, rather than from the perspective of one given field.

Once at this point, one may say: what do complex systems really have to do with an electrical grid? What does temperature variation in a honeybee hive have to do with grid frequency? Why should modelling and simulation take into account this relatively new approach? Isn't it just possible to continue using classical engineering tools for the same purpose?

Energy systems have been traditionally designed, developed and managed by electrical engineers for decades. The classical energy system is a hierarchically well-defined, top-down system, intended to work in one direction only. Centralised production using large plants for the economical generation of energy, the only possible way at the beginning of the 20th century, feed electricity into the high voltage grid, where it is transported over long distances to the origin of demand. Once there, a regional distribution grid approaches the energy to large industrial consumers and low voltage transformers for final end customers. This system worked well before deregulation of the energy markets because almost all these steps were covered by the same actor. Deregulation, along with the introduction of distributed and renewable energy sources, lead to a real paradigm shift in which the classical system was no longer sustainable. Introducing production units at lower levels firstly allowed the inversion of energy flows. Concepts like *prosumers* (consumers which are capable of producing energy as well) as well as new technologies in the field (smart metering, ICT on energy systems) etc. and especially active management of the demand side (through DSM or DR), are now leading to what is called the smart grid.

Complex systems aim to reflect the complex processes and dynamics of real systems, relying on a cross-disciplinary approach. In the course of this work we will see how the energy system is a real socio-technical system, which combines technical, automated processes with human behaviour and social factors. If we want to understand many of the processes of an energy system, only by taking into account both aspects can we draw an integrative picture.

Moreover, current methods, such as the standard profile method or other methods using low-resolution data, seem to have reached their limits. As we have seen, high-resolution models can be helpful and are necessary. Individual-centric approaches, which are a common tool in complex systems theory, can address the new challenges of using high-resolution models.

It must be said that the complex system approach does not aim to substitute any of the classical tools and approaches already existing in the field. Power flow questions, grid stability issues, and many other engineering questions have been correctly addressed and should not be represented by using overly-simplified models. However, the complex system approach aims to look at the system from a new perspective, taking it into account *as a whole* and being able to capture, for example, emergent phenomena, which can barely, if at all, be represented with other tools.

Chapter 3

Modelling and Simulation of Complex Systems

Simulation is a third way of doing science. Like deduction, it starts with a set of explicit assumptions. But unlike deduction, it does not prove theorems. Instead a simulation generates data that can be analysed inductively. Unlike typical induction, however, the simulated data comes from a rigorously specified set of rules rather than direct measurement of the real world. While induction can be used to find patterns in data, and deduction can be used to find consequences of assumptions, the purpose of simulation is to aid intuition.

Robert Axelrod [1997], on agent based simulation.

The term **modelling** has many different meanings across different disciplines. Therefore we will review what is meant by modelling. A model in the physical sense is usually some type of simplified representation of a real object. So, for example, we can have a wooden model of a car, which represents its exterior form. It has to be noted that usually the model is a simplified representation, as in this case, the wooden model will not fulfil the same operations as a real car, but rather, serves as a representation for a certain aim. For instance representing the car's shape allows for testing aerodynamics but this particular model will not be appropriate for testing the motor, for example. Furthermore, a model can simplify other aspects of the real object, as for example a scale model. It aims to represent the real entity in a reduce size, while maintaining its proportions.

Computational models can be seen as a metaphor for a physical model. Usually they implement what is called a conceptual model, which is not a concrete, physical representation but an abstract description of the reality to be represented. Through the use of computers, models can represent the real object in a virtual, computational environment, where they are implemented using specifically designed means (such as formalisms or programming

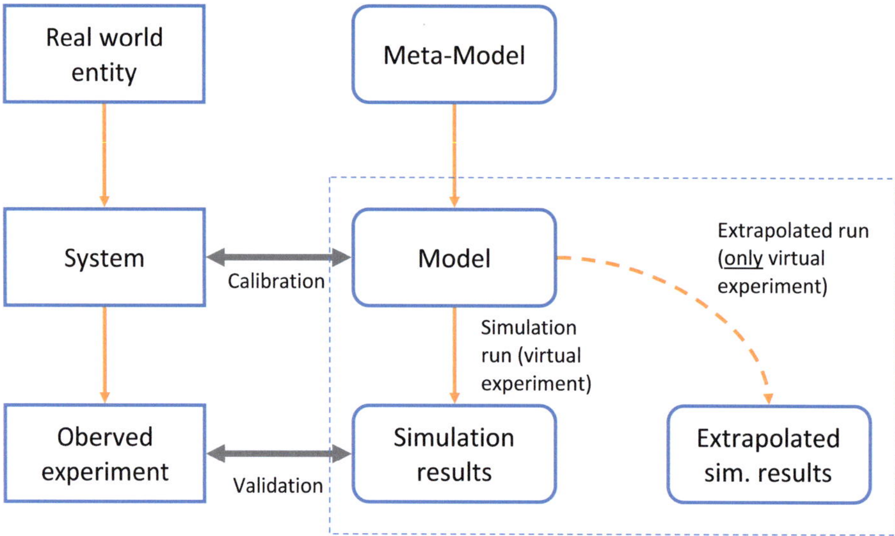

Figure 3.1: The modelling and simulation process (in the dashed rectangle) in the framework of its related real world system and an observed experiment. Adapted from Zeigler [1984]; Vangheluwe et al. [2002].

languages). This allows the performance of calculations which would be too hard to fulfil manually.

In the mathematical sense, a model is described using mathematical concepts, like equations, which describe a real system. It usually describes how certain variables are dependent on other variables and parameters, through mathematical relationships.

A rather pragmatic definition of a **model** is given by Minsky [1965]: "To an observer B, an object A' is a model of an object A to the extent that B can use A' to answer questions that interest him about A". Here, the model is used as a substitute to avoid performing tests, tasks or manipulations on the real object.

A **simulation model** is a special type of model, which is capable of evolving over time. In this context, making the model change its states over time, or simply *running* it, is called simulation. Not all models are prepared for these tasks. In our case, we aim to represent a real system as a model, in order to perform simulations on it.

There is much literature on modelling and simulation available. The modelling and simulation process is represented in Figure 3.1, which shows the aspect of a simulation as a virtual experiment. Once having created a model to represent the system to be analysed, virtual experiments are run on it. This allows the experiments to be run *in-silico*, instead of on the real system. This can have many advantages, as real experiments can be costly, risky or even impossible to perform on the real system. There is a difference between a simulation result, which can be verified against a real experiment on the system (and allows

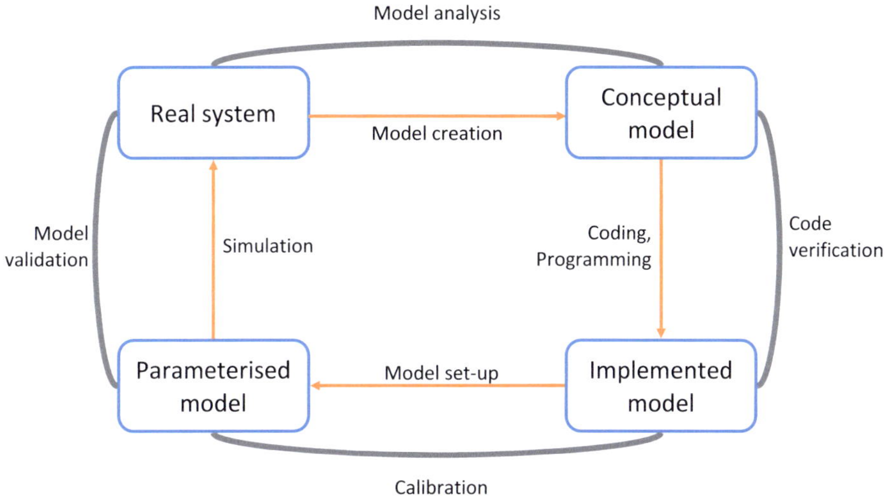

Figure 3.2: The modelling and simulation process, as an iterative process. Adapted from Refsgaard and Henriksen [2004].

us to validate the model), and an extrapolated simulation run which is based on a scenario that can't be validated in reality. The extrapolation allows the support of decision-making, by exploring future trajectories of the system. Extrapolation is any generic manipulation of the scenario or even the model itself, which modifies it from what we see in the real world. The extrapolation can be structural, spatial, temporal, etc. among others.

Refsgaard and Henriksen [2004] give an overview of the modelling and simulation process which can be seen in Figure 3.2. The main steps are: the creation of the model (or modelling), its implementation in a computational format, the setup (or scenario definition) and the simulation runs. In parallel to these steps, feedback is shown to ensure the coherence of each step of the process. Notably, the verification step ensures that the model is a correct implementation of the conceptual, or mental, representation. The final validation step compares and evaluates the simulation results against the real system. This step verifies the degree to which the model is a suitable representation of this real system.

3.1 Representation of Simulation Models

According to Burian [2010], there are two types of relationships concerning complex systems.

Structural relations: define which parts are connected together.

Functional relations: define the behaviour or dynamics of the system - how does the change of state of one part influence the state of other connected parts.

The separation of these two concepts: structural and functional, plays an important role in an initial stage of the analysis and understanding of complex systems. Furthermore, such a differentiation is advantageous when modelling and simulating complex systems. To describe a simulation model, it makes sense to distinguish between a structural and a behavioural representation.

Structural representation The structural description of the model shows only the name and type of attributes (parameters, variables) and functions (or methods) of the model. It can be seen as a black box as it defines the interfaces and internal attributes, but not the *mode of operation* of the model.

Behavioural representation The behavioural representation describes how a model evolves over time, i.e. how the variables defined above change according to given rules such as mathematical expressions, algorithms, or any other kind of relations. The behavioural description is always based on a structural description as it uses its elements as inputs and outputs (for instance, the value of a variable is updated according to the values of certain parameters by evaluating an equation).

Simulation models have different types of properties to characterise the model. Even if they both describe a value of the model at a given time, we differentiate between **variables** and **parameters**.

Variables Variables, as their name indicates, are able to vary their value throughout the simulation. They are usually used for outputs or intermediate results or states. Their values can be read or evaluated, but are usually not changed from outside the model.

Parameters On the other hand, parameter values are fixed for an entire simulation. A parameter is used to describe a certain state that is intentionally fixed before running the simulation. It is usually constant throughout. However, the value can differ from simulation to simulation, for example through different instantiations of the same model.

Scenarios A set of parameters describes the characteristics of a simulation run, which we call a scenario. A scenario is a specific, initial configuration of the model. We can further expand a scenario which will also describe specific actions. A scenario can describe events that might affect the model-run at a given time. These can be discrete events, such as a sudden change of value or state, induced intentionally (which are not a result of the model behaviour itself). A continuous variation of a value can be part of the scenario description, too, such as the evolution of a temperature or a population over time.

3.2 Agent Systems, Models and Simulations

Agent systems have been an emerging field of research over the past few decades. Agent systems are closely related to the study of complex systems. They have been applied in several disciplines, in some of which they are already well established (for example sociology or biology, in relation to behavioural studies). It is challenging to identify a common definition for these systems, so this chapter will present the different terms and definitions that can be found, and will highlight the differences between them. In the first instance, **multi-agent systems** (MAS) will be discussed to introduce **agent-based models** (ABM). It is not usual to find both terms used at the same time in literature. Many definitions are available and they will be explored in the following section. It is interesting to note, that it is difficult to find a comparison of the terms MAS and ABM. The distinction between MAS and ABM is very fine but this work does distinguish between them.

It is generally accepted that agent systems are well suited as a tool to model and simulate different types of complex systems. They allow us to represent many of the dynamics that arise in those systems through simulation, and to study their effects in a virtual environment.

3.2.1 Multi-Agent Systems

Identified as the most generic of the concepts, multi-agent systems (MAS), are generally systems that are composed of multiple agents, with varying degrees of intelligence, that interact with each other in some way.

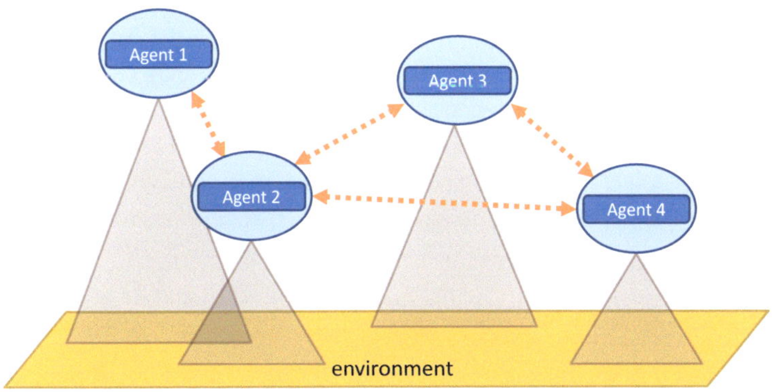

Figure 3.3: A multi-agent system: the agents present some relations between them (orange) and have a limited perception of the environment.

Ferber [1999] defines the term MAS based on a system which comprises the following elements:

1. An environment, that is a space which generally has a volume.

2. A set of objects. These objects are situated, that is to say, it is possible, at a given moment to associate any object with a position in the environment. These objects are passive, that is, they can be perceived, created, destroyed and modified by the agents.
3. An assembly of agents, which are specific objects, representing the active entities of the system.
4. An assembly of relations, which link objects (and thus agents) to each other.
5. An assembly of operations, making it possible for the agents to perceive, produce, consume, transform and manipulate objects.
6. Operators with the task of representing the application of these operations and the reaction of the world to this attempt at modification, which we shall call the laws of the universe.

These characteristics show that MAS are directly related to complex systems, by sharing many of their properties. It should be noted that, when referring to a MAS, we are referring to the system, independently whether a real or a virtual system is meant. Furthermore, the modelling aspect is not yet considered at this stage. The term *multi* indicates that here multiple agents are considered in the system, usually a large number of them.

3.2.2 Agent-Based Models

Agent-based models(ABM) (also known as multi-agent models or individual based models) are the foundation of a software development approach for the simulation of complex systems. They are independent of the area of knowledge that they refer to. Agent-based models have been increasingly used over the past few years in, and for systems which are composed of large number of units that take decisions autonomously, and are therefore, heterogeneous in nature and unpredictable in essence. These systems are generally called complex systems, although this is not a universally agreed term throughout different disciplines.

The following definitions can be found in literature:

- “Agent-based modeling (ABM) is a way to model the dynamics of complex systems and complex adaptive systems. Such systems often self-organize themselves and create emergent order. Agent-based models also include models of behavior (human or otherwise) and are used to observe the collective effects of agent behaviors and interactions. The development of agent modeling tools, the availability of micro-data, and advances in computation have made possible a growing number of agent-based applications across a variety of domains and disciplines.” [Macal and North, 2010]

- "The ABM approach consists of a decentralized collection of agents acting autonomously in various contexts. The massively parallel and local interactions can give rise to path dependencies, dynamic returns and their interaction. In such an environment global phenomena such as the development and diffusion of technologies, the emergence of networks, herd-behavior etc. which cause the transformation of the observed system can be modeled adequately. This modeling approach focuses on depicting the agents, their relationships and the processes governing the transformation." [Pyka and Grebel, 2006]
- "Formally, agent-based modeling is a computational method that enables a researcher to create, analyze, and experiment with models composed of agents that interact within an environment." [Gilbert, 2007]

Basically, ABM focuses on the modeling of systems at a local level through the definition of their elementary units (called agents), as well as through their interactions. In this sense, it can be said that ABM is a form of modelling MAS. The elementary units are intended to be modelled in a simple way, while the complexity of the system is an emergent property of their interactions. There are three main groups of actions that must be modelled:

1. Sensing the environment: Agents are capable of acquiring information on the local environment through sensors.
2. Taking decisions: Each agent can autonomously decide what action should be taken regarding its local information in order to fulfill its objectives.
3. Reaction to the environment: Agents decisions are effected on the environment through actuators. Therefore, a feedback loop exists between the environment and the agents.

Because agents are defined through local information, an important difference between them and other models is their heterogeneity. Each agent is modelled in the same way, but the input parameters and the local environment can be different. Therefore, decisions taken by the agents can differ, even when the reasoning process and objectives remain the same. This is an important difference from traditional models, where several objects of the same type usually share the same inputs.

Through the agents interaction, emergent phenomena can appear at higher levels. These emergent phenomena are not necessarily predictable from their behaviour at an individual level. Typical examples in engineering systems are synchronization or de-synchronization processes. Another emergent behavior that is a focus of research through ABM is self-organization. Traditional models are not able to reproduce emergent processes and therefore they are predictable, from the modeling stage.

ABM intends to define the individual agents in a simple way, in order to be massively and easily replicated. The structure of the system can therefore be dynamically changed through the addition or elimination of individual agents, even in simulation time. It is possible, therefore, to represent long term scenarios, where the structure of the system is in the process of continuous change.

3.2.3 MAS or ABM?

There are no generally accepted definitions for MAS or ABM. Though, we can note some differences between these terms. These differences seem to be common according to the majority of the definitions and scope given above.

- MAS refers to a system, where many agents are interacting with each another, in some way, and mostly communicating. Those systems usually are related to large numbers of agents in an environment. They can be real or virtual.
- ABM refers to a modeling paradigm (class of computational model) which is based on "the theory of" agents. It can be certainly used to model a MAS. It refers rather to the way of modeling entities as agents, giving them the capability to interact.
- ABM is usually conceived in order to perform simulations (as a simulation model). Here, usually the terms agent based simulation (ABS) can be found. ABS are simulations performed by ABMs. A MAS is not exclusively a simulation model, but can be used as one.

In different languages, terms are being literally translated and used different ways. In France, for example, it is common to talk about *Systèmes Multi-Agent* referring in general to agent-based systems, models and often also to simulation. In English, multi-agent system is not directly linked to the modelling and simulation fields. It is used, rather, for a system that makes use of autonomous, distributed decision making processes.

3.2.4 The Agent: a Definition

Agents are the most important part of MAS and ABMs, especially when in relation to other, not individual-centric approaches. A commonly accepted definition of an agent is difficult to find. The most generally accepted definitions are from Weiss and Wooldridge.

Weiss [1999] states that "agents are autonomous, computational entities that can be viewed as perceiving their environment through sensors and acting upon their environment through its effectors".

For Wooldridge [2009], "an agent is a computer system that is situated in some environment, and that is capable of autonomous action in this environment in order to meet its

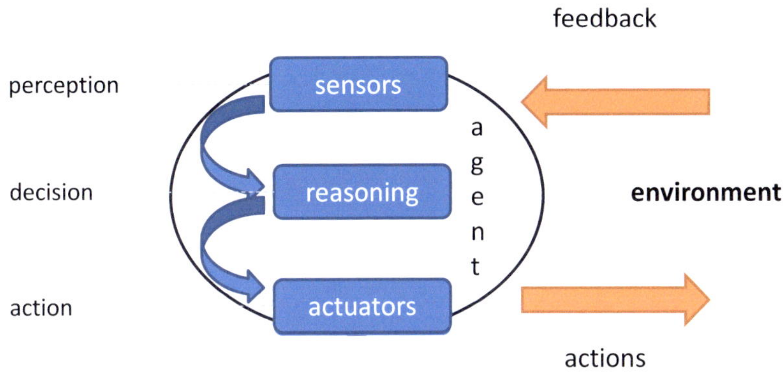

Figure 3.4: Structure of a generic agent. Adapted from Wooldridge [2009].

delegated objectives". This definition can be generalized by defining agents as any type of systems, thus not restricting the definition to *computer* systems.

This aspect becomes already clear in Ferber's [1995] definition, which defines that "an agent is a real or virtual entity whose behavior is autonomous, operating in an environment that is capable of perceiving and which it is able to act, and interact with other agents".

An agent is a physical or virtual entity

- which is capable of acting in an environment,
- which can communicate directly with other agents,
- which is driven by a set of tendencies (in the form of individual objectives or of a satisfaction/survival function which it tries to optimise,
- which possesses resources of its own,
- which is capable of perceiving its environment (but to a limited extent),
- which has only a partial representation of this environment (or, perhaps, none at all),
- which possesses skills and can offer services,
- which may be able to reproduce itself,
- whose behaviour tends towards satisfying its objectives, taking account of the resources and skills available to it and, depending on its perception, its representations and the communications it receives.

The definition includes several significant aspects such as, for example, that the agent can be a physical entity, such as a robot or a machine, which is acting in the real world. However, it can be also a completely virtual entity, such as a computer program that does not have any physical existence at all and does not need to interact with the real world.

Characteristics of the Agent

Ferber and Perrot [1995] believe that autonomy is the most fundamental characteristic of the agent. Agents seek to meet individual goals in order to jointly achieve a particular objective that their system should achieve. Autonomy allows the agent to answer or ignore requests from other agents. Each agent therefore has some degree of freedom to act, which differentiates it from other similar concepts, such as objects, or processes. The agent is thus both an open system because it needs external elements, and a closed system because it organises itself. Agents, like classical AI systems are capable of reasoning. They can also, however, perform actions which can alter their environment. The agents' environment can be defined as everything (that is not an agent) but that interacts with the agent. Agents can also communicate with each other and this is one of the main modes of interaction between agents [Demazeau and Costa, 1996; Abras, 2009].

In order to call something an *agent*, it should at least fulfill some basic properties: these are autonomy in its actions (the decision for acting is taken by the agent himself and not determined from the exterior), and the agent is able to interact in some way with its environment or with other agents. Moreover, an agent disposes only on local information, which is coupled in some way with the environment. Further properties such as pro-activity or intelligence are typical for agents, but not obligatory. What is important is that the agent has been conceived with the intention of being able to fulfil these extended functions. It is important to enlarge the definition in this way, as it permits a representation of large *real* objects as agents, event if they only fulfil rather simple tasks.

3.2.5 The Environment and its Interaction with the Agents

The environment is defined as the part of the system within which the agent operates. It is not the agent itself nor is it any of the other agents, but rather it is everything that has an (external) influence upon it. As visualised in Figure 3.5, different types of environments can be identified. They can be simple, multi-layered, or even change over time. Certainly, an environment can be sufficiently described by one single layer. Even complex stimuli can, under certain conditions, be aggregated into one single stimulus to causes one certain reaction. Still, there are some situations where different stimuli cannot be satisfactorily combined, so that the independent layers for an environment need to be defined.

3.2.6 Agent Networks and Social Ability

While the environment concerns itself with reactivity, networks are related to social ability. Agent networks are made up solely of other agents, although they need not be the same kind as the originating agent. Also, in a similar manner as the environment, there might be more than one network to which the agent is connected. Likewise, the networks might change over time, with nodes accruing or dropping out (even to the extreme of having a complete layer, or even all layers, disappear completely). Figure 3.6 shows different possible network configurations. A network can be classified by layer (single- or multi-layer) and connection type (static or dynamic).

Agents can form complex network topologies, such as scale free or small world configurations, as discussed in Section 2.2.2.

3.2.7 Proactivity

We have now come to associate the terms *reactivity* with the environment within which agents embedded and *social ability* with a network to which an agent is connected. The next step is to define a motivation within the agent, so as to allow it to make use of its surroundings.

The concept of proactivity, introduces the idea that an agent follows a specific goal or pursues a task. With this, the aim is to signify that each agent has a specific goal, usually defined as a set of objectives, that essentially justify the reason for the existence of the agent. For the most part, these objectives are preprogrammed into the agent, though it need not be so. Objectives can be concrete (e. g., a mathematical operation to perform when certain conditions apply) or very abstract (e. g., *thrive and prosper*.) Abstract objectives would need to be interpreted and broken into smaller steps to be achieved separately,

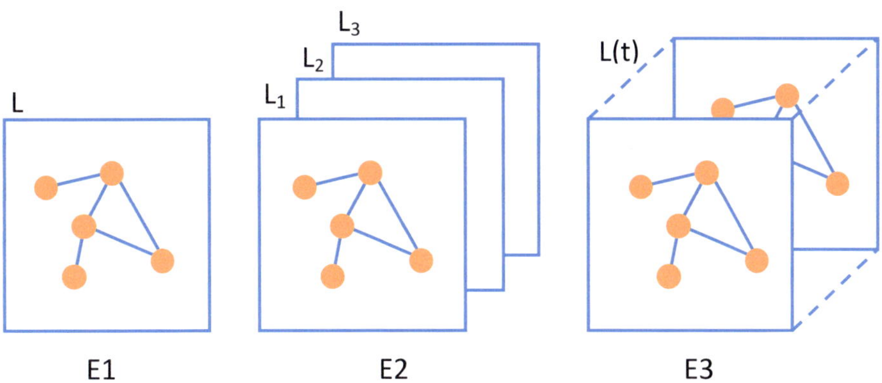

Figure 3.5: Different environments: E1: single layer, E2: multi-layer, E3: continuous changing environment.

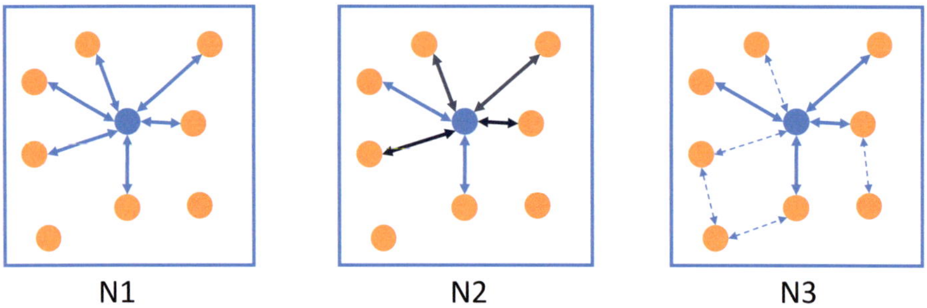

Figure 3.6: Network types: N1: single layer, static network N2: multi-layer, static network N3: single-layer, dynamic network.

in order to to fulfil the final objective). There are numerous strategies to achieve these objectives.

Even before that, a system must be in place through which to define a problem to solve (the objective) and the logic to be able to successively solve it. Automated problem solving can be quite complex. In any case, a decision-making strategy (or algorithm) is essential. Usually, some sort of evaluating function (often also called *utility function*) needs to be applied in order to assess whether the planned task (or, more probably, the planned sub-task) fulfils certain criteria most notably, if it is a step in the *right direction*.

It is important to understand why the agent has to be provided with the ability to make decisions. It is a direct requirement for the agent to fulfil its own goals. It should be noted, that within a system or model containing more than one agent, each agent has its own goal. The goals may be similar or even identical, however, each agent follows its own strategy and subsequent actions in order to pursue that goal. Taking into consideration the differences in how local environmental characteristics and network setups differ from one agent to the next, it is this heterogeneity that enables the agents to show very different behavior. As the only difference is outside the agent, it follows that an agent's behavior differs according to that same agent's surroundings. When that thought is followed, it is fairly easy to see that emergence can appear when a number of agents start working collectively with each other while trying to achieve its own, separate goal. In the field of social sciences or, specifically, sociology, such behavioural patterns are a matter of great interest. This is exactly where agent-based systems gain their potential from. [Apicella, 2009]

3.2.8 Agent-Based Simulations

Agent based models are commonly simulation models (see Section 3.1). Therefore, the terms Agent-Based Simulation (ABS), which normally refer to a simulation performed with an ABM, or Agent-Based Simulation Model (ABSM) are used in this context. Different parameterisation and initial conditions can generate specific scenario configurations of the ABM. These scenarios generate each a set of simulations, which are subsets of the total population of simulations [Chen et al., 2008]. Figure 3.7 shows these concepts.

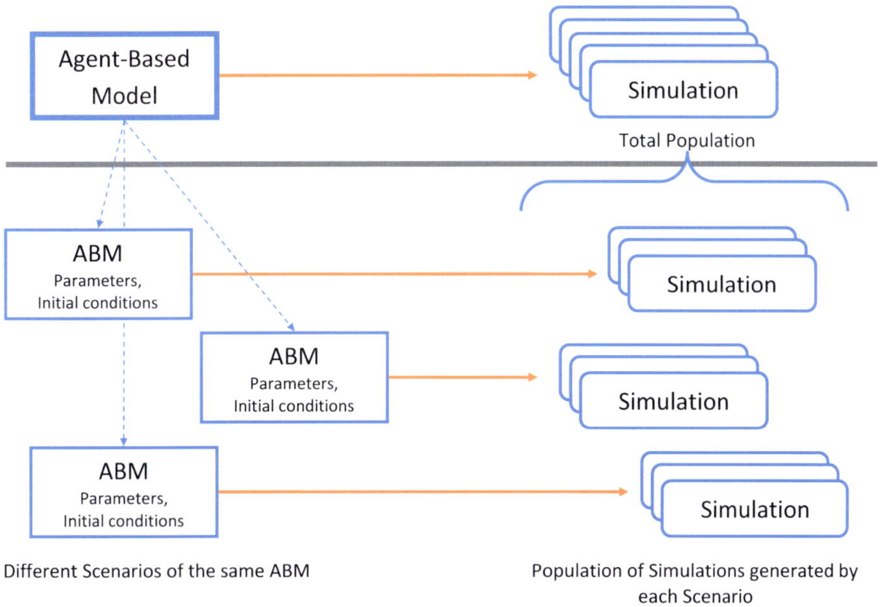

Figure 3.7: The ABM generating simulations, different scenarios due to parametrisation and generation of scenario simulations. Adapted from [Chen et al., 2008]

3.2.9 Agents and Objects

Object oriented programming (OOP) is a programming paradigm which focuses more on *objects* rather than on *actions*. Before introducing OOP, a program was seen as a logical procedure which processes data to create results. OOP veered away from this point of view, placing more importance on how to define the data than to programme the logic. The objects are defined through their relationship. Abstraction provides classification and organization for data and objects. For this purpose, OOP now introduces widely spread concepts such as generalisations (inheritance) or handling different data types using the same interface (polymorphism).

Agent-based systems share many properties with the objects defined in the OOP paradigm. Encapsulation is a fundamental principle that describes a restriction of access to some of the components of an object, and is applied for agent systems in a similar way. Separation of concerns describes the process of separating software into distinct features that overlap in their functionality as little as possible. In agent based systems, an agent has specific objectives and for different concerns usually different agents are used.

Object oriented programming (OOP) exists as a formalism however, the agent-based system paradigm does not yet exist, and is not yet as well defined as OOP. There is a certain tendency that agents somehow represent more *intelligent* objects. According to our definitions given in Section 3.2.4, an important point is that agents have a proactive role on clearly defined objectives. These aspects, combined with a certain amount of autonomy, differ from the more static objects. Classes in OOP usually have clearly defined goals, but there is no need for a proactive role.

Wooldridge [2009] sums up the differences in a single phrase:

> *Objects do it for free, agents do it because they want to.*

Interactions between agents are usually more complex than between objects. Message passing is a good example of this difference. In OOP, methods are usually called on the objects, which perform some operations and are pre-programmed to return a given result (usually defined as a result type, such as a primitive type (double, integer) or a type of object). Message passing, which is used to communicate between agents, is a more *natural* way of communicating, in the sense that messages are sent to a recipient. So far, this is similar to a method call. However, when a message is sent, the recipient is the one responsible for reacting to it, and can decide locally what action to take. In the case of a method, a predefined code will be run, and is implicitly considered in the method call. Sending messages also allows for communication mechanisms like broadcasting, where a message is sent to the whole network, and only the agents which feel concerned will react to it, depending on their local criteria, rules or state.

Message passing, from the implementation point of view, is also a more complex communication method. However, it clearly offers more flexibility and is a better adapted representation of reality in modelling a real system.

3.2.10 Drawbacks and Benefits of ABM

Although there are many examples of applied ABMs, only few evaluations of the method can be found. Usually the applications are not compared to other approaches, and the ABM community tends to be quite enthusiastic in its approach and avoids confrontation with other methods. However, the recognition of the method in other areas is a crucial point in establishing the foundations of a generally accepted theory on agent systems.

Jennings [2000] states that *against this background, there are two major drawbacks associated with the very essence of an agent-based approach: the patterns and the outcomes of the interactions are inherently unpredictable; predicting the behavior of the overall system based on its constituent components is extremely difficult (sometimes impossible) because of the strong possibility of emergent behavior.*

In the course of this thesis, there has been an effort to identify relevant drawbacks and benefits of ABM.

Drawbacks Firstly, due to their individual-centric approach, in which agents can represent large numbers of real entities, a large number of data is necessary to parametrise these models. Also, large datasets are available as output of the models, and have to be correctly handled and interpreted. This involves voluminous databases on input and output. For input, the existence of large input datasets is presupposed. This is one of the major drawbacks, as in many cases the data is unavailable. However, methods to generate synthetic data based on real populations are in development [Thiriot and Kant, 2008].

The calibration and validation of these models is difficult, as we deal with a large number of entities, in most cases, which cannot be individually verified. Statistical methods can be used to assure accuracy at a statistically representative level, however they do not guarantee correctness at the individual level. Therefore, different calibration and validation phases are needed, usually at different levels. ABMs can grow rapidly thus need large computational power, as they are scalable and usually large numbers of agents have to be simulated along with their interactions. Parallelising these types of models is also a rather hard task, due to the large number of interactions. Moreover, especially due to their non-determinism, ABM results have to be correctly interpreted in order to identify and understand, for instance, emergent phenomena.. Finally, because of their ability to simulate unexpected behaviours, it is difficult to distinguish between *bugs* and the desired behaviour.

Benefits The modelling approach is based on simple, but heterogeneous, entities (with heterogeneous, autonomous behaviours), which allow a detailed and more realistic representation of complex systems. ABMs are capable of evolving over time, so it is possible to add or remove agents and even change the structure of the system during the simulation. Agents are represented in a simplified but intuitive way, as they are described by real entities of the system, often located in a concrete environment. ABMs provide integrative solutions that embed many simplified models in order to represent a complex system with a large number of components and interactions. The visualisation and representation of the simulated phenomena can significantly improve exploration, discussion and information exchange on the models. ABMs allow the extraction of high resolution data, even for mid- and long term scenarios. Because of their stochastic, non-deterministic nature, results can be given in the form of solution spaces instead of deterministic results (for example using methods like Monte-Carlo). An agent-based approach allows us to represent, in a more intuitive way, the relationship between causes and effects. As we will see in the case studies presented in the previous chapters, ABMs allow us to better understand the behaviour of complex systems and their phenomena, through an exploration of the modelled system. We will see how an ABM helps us to understand complex, emergent phenomena and explain their causes at an individual level.

3.3 System Dynamics

System Dynamics (SD) is a method which was developed by J. Forrester in the 1950's at the Massachusetts Institute of Technology [Forrester, 1961]. It is intended to analyse and simulate (through modelling) complex and dynamic systems at system level. It is based on a representation using stocks and flows and characterized by using internal feedback loops. These loops affect the behavior of the entire system, as they are able to reproduce the non-linearities of complex systems.

System dynamics falls under the so-called macro-simulation approaches, which describe the behavior at a high, aggregated system level. SD is applied where a behavior of individual entities is not directly regarded and does not have to be differentiated. However, knowledge at the aggregated level must be available. Meaningful variables defining the system behavior must be identified. They are then related to each other through differential equations.

Being originally developed to understand the complexity in industrial processes, SD is now applied to a wide variety of fields. Most of the applications are related to policy making and analysis. Qualitative approaches and models are common. However, the many *soft* variables in the models, which sometimes make it difficult to estimate parameters for a quantitative application, have resulted in this kind of application being less well established.

The simulation of SD models is run computationally on small step calculations (numerical approach solving methods) that update all the variables at each time step. These methods are needed to correctly represent the feedback effects, as the resulting mathematical systems which need to be solved are also usually of a non-linear nature. A SD model addresses the values of the state of the system, which varies over time. SD defines several formalisms in order to represent its models. There are two types of diagrams: casual loop diagrams and stock and flow diagrams.

3.3.1 Casual Loop Diagrams

Casual Loop Diagrams (CLD) show the casual relationships of the system in a graphic way. They help visualise the causes and effects that appear among the interrelated variables of the system. The CLD diagram is constructed by connecting different nodes. It is a directed graph with nodes representing system variables and directed arcs representing the binary relationships between these variables.

We can distinguish between two types of feedback loops:

- Positive reinforcement: creates a regeneration of the effect, by feeding back part of the output which increases input. They are also called reinforcement loops.
- Negative reinforcement: creates a degeneration of the effect, by feeding back part of the output which is opposed to the input and reduces it. They are also called balancing loops.

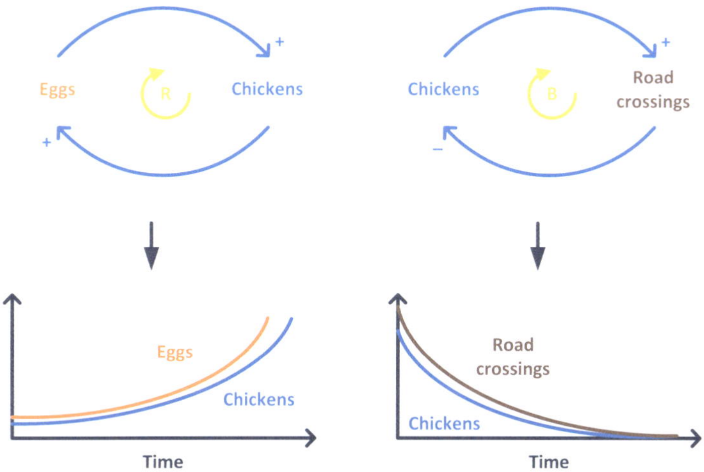

(a) Reinforcement loop **(b)** Balancing loop

Figure 3.8: Casual loop diagrams in system dynamics: (a) shows an example of a reinforcement loop and its corresponding dynamics, which cause an exponential growth. In (b), a balancing loop can be seen, which results in an exponential decay in its behaviour over time.

3.3.2 Stock and Flow Diagrams

Whereas casual loop diagrams are used for conceptual development of models, a second formalism, which is closely related to CLD, forms part of the SD paradigm. Stock and flow diagrams are quantitative and can be used for simulation.

Stocks represent a quantity measured at a specific a time. A flow variable however, is measured over an interval of time and is measured in units per time. A simple example to illustrate this is a water storage tank with an inflow and an outflow. The water level can be measured at a specific time (stock). To quantify the flows, we have to determine how much volume per unit of time is flowing in and out to the tank (flows).

In the mathematical sense, the relation of stocks and flows is given by the derivative. Stocks integrate all of the incoming or outgoing flows:

$$Stock = C - \int_0^t sales(\tau)d\tau \tag{3.1}$$

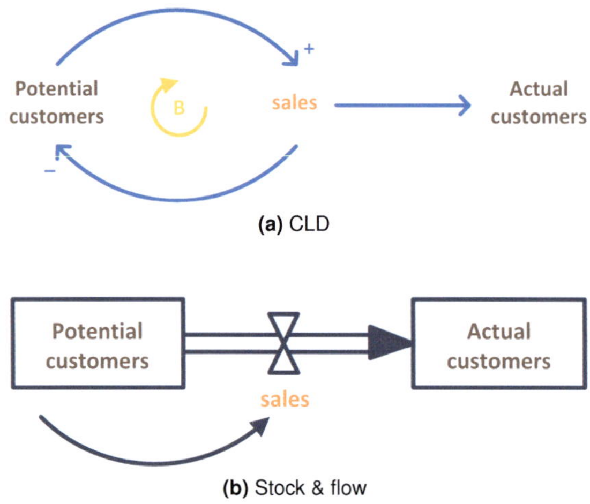

Figure 3.9: Casual loop diagram and stock and flow diagram: (a) shows an example of a casual loop diagram. In the transformation into a stock and flow diagram, it can be seen that two variables correspond to a *stock*, and one variable to a *flow*.

where C is the initial value of the stock.

3.3.3 When to use System Dynamics

When comparing with ABM, SD is rather a top-level approach. This makes it suitable for representing systems in which the aggregated behavior must be studied. It allows for the representation of complex processes, such as feedback loops, which can be represented in an intuitive way by identifying them through the use of CLD. Furthermore, system dynamics can be used to represent almost every system that can be described using ordinary differential equations (ODE), as will be shown later. This makes SD a powerful approach for energy system modelling. If we want to disaggregate that approach, individual components can be modelled by taking only a proportional part of the SD model, usually not a realistic approach, or by representing the individual through agent based models. One of the main differences between SD and ABM is that SD is always a continuous approach (ODEs are solved for a continuous time horizon), whereas ABM is usually based on rather discrete event models. As we will see in the next section, ABM can also integrate continuous models through a multi-method approach.

3.4 Discrete Event Modelling

In a discrete event model, system models are described at an abstraction level where the time base is continuous, but during a bounded time-span, only a finite number of relevant events occur. The state of the system can change due to these events. In between events, the state of the system is not allowed to change. This is unlike the continuous models in which the state of the system may change continuously over time Vangheluwe et al. [2002].

One of the formalisms which is based on discrete events is the state-chart. State-charts were firstly unified by Harel [1987], who proposed a notation which is now generally accepted throughout different fields. For example, it is the base for the normalisation of state-charts in the Unified Modelling Language (UML). State-charts are representations of finite state automata (as they are known in computer science). This formalism allows the representation of the complexity of these automata in a manageable way.

3.5 Multi-Approach Modelling

Models of system behaviour can be represented at different levels of abstraction or detail as well as through different modelling methods and formalisms. The particular method and level of abstraction depends on the background and goals of the modeller, as well as on the system which is to be modelled [Vangheluwe et al., 2002].

Decision-making for a complex system requires an understanding of the behaviour of the overall system. Studying the individual components of the system will usually not be enough. The complexity of this system and its model is due to several factors.

- The number of interacting, coupled, concurrent components.
- The variety of component formalisms. Often, a mix of continuous and discrete components occurs.
- The variety of views, at different levels of abstraction.

Complex behaviour is often a consequence of a large number of interactions. These interactions can also create feedback loops which reinforce this complex behaviour. In order to represent these aspects, one formalism or modelling paradigm is usually not enough, as it will reduce the point of view to a certain level of abstraction or to a specific component; or will neglect the interactions between them. Also, depending on the questions to be answered by the model, different abstraction levels have to be chosen. Components of the system itself may require different levels of abstractions, depending on their role in the system. The choice of the modellings paradigm is usually related to:

- the results that need to be provided by the model,
- the abstraction level,
- the amount and quality of data available for calibration and validation and

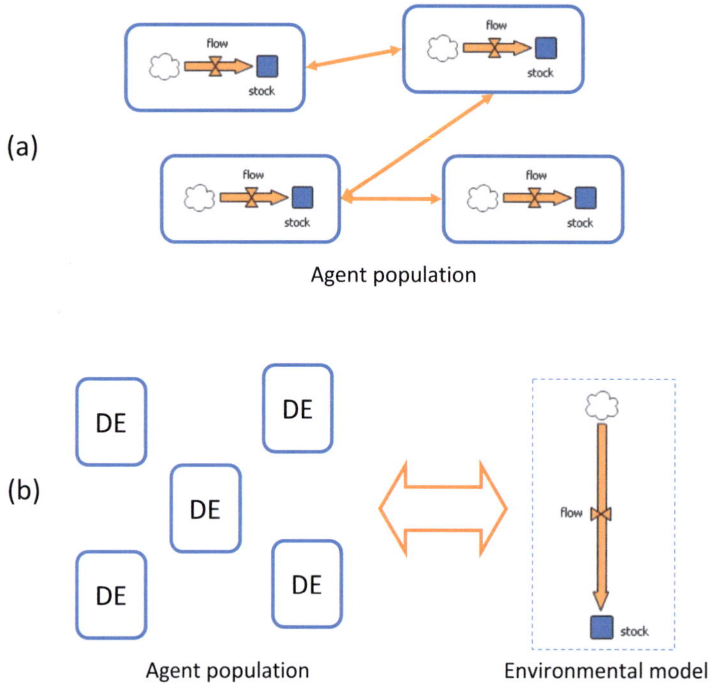

Figure 3.10: Examples of different possible architectures for multi-approach models. In (a), agents contain a behaviour which is modelled using SD. Below, in (b), agents are modelled using DE and interact with an SD model at the population level.

- the availability of tools and solvers for the given paradigm [Vangheluwe et al., 2002].

In [Borshchev and Filippov, 2004] three major approaches are compared and analysed to determine their affinity to multi-method modelling. The approaches suggested are agent-based modelling, system dynamics and discrete event modelling. ABM seems able to capture more real life phenomena than SD or DE, however ABM cannot replace the other approaches in all cases. In many cases, SD or DE models can offer correct answers in a much more efficient way. Wherever this is the case, the traditional approaches should be used. Although, the use of different modelling paradigms should be considered when the system is composed of different parts which can be individually modelled with different approaches. Different modelling architectures for multi-approach systems are possible (see Figure 3.10).

3.6 Systems-of-Systems

The term system-of-systems (SoS) is a collection of task-oriented or dedicated systems that pool their resources and capabilities together to achieve a new, more complex meta-system. This meta-system itself offers more functionality and performance than simply the sum of the individual systems from which it is constituted [Markarian et al., 2011].

One can easily argue [Kremers et al., 2010] that energy systems, and especially smart energy systems, which tend to be more distributed and decentralised, mostly satisfy the five principal characteristics that distinguish them as true systems-of-systems, as defined by Maier [1998]:

1. Operational independence of the elements: The elements of the energy systems are able to operate independently and usefully while fulfilling autonomous goals. Producers, consumers and other elements can work properly when they are not connected to the energy system as a whole, or can work as independent and autonomous systems themselves (for example microgrids).

2. Managerial independence of the elements: energy systems not only *can* operate independently, they *do* operate independently. They maintain a continuing operational existence independent of the SoS.

3. Evolutionary development: The energy system does not appear fully formed. Its future development and existence is evolutionary and depends of other environmental agents.

4. Emergent behaviour: The energy system performs functions and carries out purposes that do not reside in any component system. These behaviours are emergent properties of the entire SoS and cannot be localised to any component system. The main goals of the electrical system are given by these behaviours.

5. Geographic distribution: The geographic extent of the energy system is large. The components of the energy system can readily exchange information (and, in this case, can also exchange energy).

3.7 Calibration and Validation of Agent-Based Models

Calibration describes the process of adjusting the model and its parameters in order to obtain desired results. A validation is usually an evaluation of those results through comparison with real data. The validation of agent-based models is, itself, a complex subject. Scepticism about ABM is because the validation process is not easy and cannot be performed in the same way as in deterministic models. As agent-based models usually cover different scales of the system (at least two, the individual and the systemic one), a validation should be performed at these two scales, at a minimum. An individual validation is possible by the comparison of individual agents with real data or measurements. In the case of energy systems, an individual power plant or a specific consumer can be parameterized and compared with real data. The validity of an individual model can therefore be confirmed, as in classical validations.

However, when aggregating a population of replicated agents, the results at population level are also relevant (usually more relevant than the individual ones). Once having calibrated the individual agents, a calibration of the model at an aggregated level should take place. This is usually possible through the available macro-data (in the case of energy systems, measurements at an aggregated scale such as a transformer or substation).

This second validation has to performed in consistence with the first one. Aggregated validations are closely related to the statistical effects of the system. Heterogeneity among different agents can impact the stability of the results. Therefore, probabilistic methods such as Monte-Carlo simulations should be used. However, a second validation at system level does not ensure, that the representation of the system at intermediate levels will be correct. In only confirms that, at an aggregated level, the results are correct. It does not reveal anything about intermediate aggregations of parts of the population or effects in local clusters. In either case, a two step validation already ensures consistency between two levels of the system. This might suggest that intermediate levels will not suffer a too large degree of incorrectness. If enough data is available, additional calibrations can be performed. These will, however, increase the degree of complexity of the process, as consistence between all the calibrated levels should be maintained. Furthermore, if dealing with multi-scale models, a validation at each main scale should be performed at a minimum.

It is unlikely that a complete validation of an agent-based model could be achieved because of the large amount of data and the complexity of the model itself. Additional expert validations are very important in order to ensure the validity of the models.

3.8 Complex System Modelling – also in the Energy Field?

Once we have given an overview of modelling and simulation of complex systems in general, one may ask what this approach can add to the modelling of complex systems. The complex modelling approach does not aim to substitute current modelling tools, but rather model the system from a new point of view. In order to model a system such as the electrical energy system using a complex approach, agent-based modelling seems a reasonable representation of the system from an individual perspective. This considers its different, single components. Furthermore, it could be interesting to include non-technical aspects into the model, as the complexity approach is inherently interdisciplinary and considers not only one domain. This could enrich the models by including human behavioural or environmental aspects such as meteorology. Of course, this is not an easy task. Many questions arise, such as:

- How can we be sure that an individual model of an entity will be correct?
- How can we represent different objects of the same entity in a realistic way (think about TV, which might be the same concept but have many different technological implementations (cathod ray, LCD, plasma, etc.)?
- How do we know that a population of individual models correctly represents the real population?
- Is it worth the effort to model individually when averaged models already give good answers?
- And last but not least: What is the real added value of this approach?

These questions will be addressed in the following chapters. At this stage, it is clear that the increasing complexity of the electrical system, such as the trend towards a smart grid, calls many traditional methods into question. The classical approaches are not able to handle systemic issues however, modelling the system from a complex point of view can possibly overcome some of these limitations. The need for coupling technical and social behavior models seems to be essential to capture systemic behaviours in complex infrastructure systems as the electrical grid.

Chapter 4

Electrical Energy System Modelling

> *It can scarcely be denied that the supreme goal of all theory is to make the irreducible basic elements as simple and as few as possible without having to surrender the adequate representation of a single datum of experience.*
>
> Albert Einstein [1934].

In order to represent an energy system as a complex system through an agent-based approach, it is necessary to individually model the components. Therefore, a special focus is put on modelling the components, or entities of the system, as they behave rather than taking a macro-system approach of using statistical models or averaged, high level representations. Entities of the electrical system need to be modelled by taking into account their individual states and modes of operation. This can also be called micro-modelling, as for example, the objective is not to model a consumer by statistical means, representing an aggregated, averaged and typical behaviour. Rather, each individual household should be modelled, parameterized by information which can be taken from statistical datasets, and then recreating each household as an individual entity generating a power curve.

It is important to note that a very deep and detailed approach is an ambitious challenge. As Albert Einstein's quotation indicates, the models should be made as simple as possible, but no simpler. There are already a large number of energy system models, each very specific and detailed for their concrete area of application. This is the case for classical engineering models. which represent the technical aspects of the power grid. These could be the representation of the physical processes of the system, such as electrical, magnetic or charge effects, by using the laws of Ohm, Joule, Gauss, or the Maxwell equations. Of course, for technical analysis, these forms of representation are very important. However, when representing the energy system as a whole by modelling its individual components, the representation of a large number of entities in this way can consume a lot of resources. Possibly, depending on the questions to be solved, this might not even be needed. So, a way has to be found to simplify, or to compromise on which of the technical parts of the system will be modelled and to what degree of detail. The approach taken in this work aims to extend this rather technical view of the system by modelling other system-relevant areas. These can be social behaviours (the actions of humans which have a direct or indirect influence on the energy system) or the inclusion of communication flows or other type of interactions that do not reside purely on the power grid level. Increasing amounts of interactions with the environment (meteorological phenomena like wind or solar radiation), through the inclusion of RES, have an effect that cannot be ignored.

In a first stage, a review of existing models in the energy domain will be given. The agent-based approach is quite new to this field, but some applications related to energy have been made over the last few years,. This work makes the effort to better understand the relationship between the complex electrical system and modelling and simulation methods of generic complex systems.

4.1 Agent-Based Models for Energy Systems: A Review

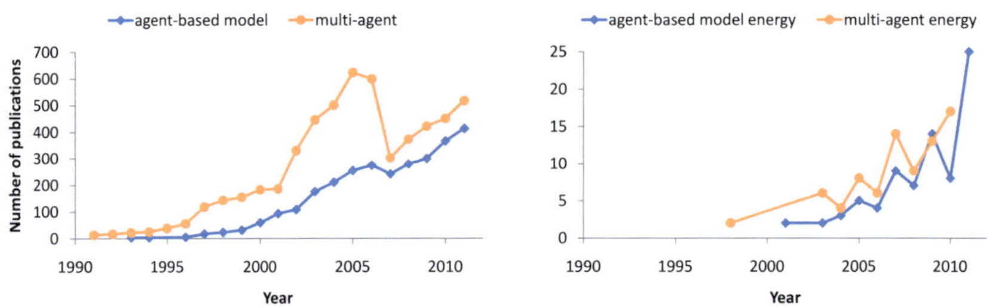

Figure 4.1: Number of publications of selected keywords found on the ISI Web of Knowledge for the period 1990 - 2012. On the left, for general agent systems; on the right, for agent systems related to energy[1].

A general search on the terms *agent-based model* and *multi-agent* shows a strongly increasing number of publications which match these keywords. Hardly any publications can be found dating from before 1990. When adding the term *energy* to the search, a low number of publications is found, which denotes the innovative character of using this approach in the energy sector.

Models using the agent-based paradigm are currently state-of-art in a few fields of the energy sector. In most cases, agents are representations of market actors dealing with the problem of market liberalisation [Sensfuss, 2007]. For several years it has been attempted to improve economic models by the introduction of heterogeneous behaviours into the more static and homogeneous, traditional modelling techniques [Ríos-Rull, 1995]. Some

[1]Energy is taken as a general keyword. Even if most of the publications were related to (electrical) energy systems, energy as it is understood in biology, physics and other fields can also be applied.

agent-based models like Sugarscape by [Epstein and Axtell, 1996; Epstein, 2006] show the emergence of counter-intuitive behaviours related to distributed decision-making from heterogeneous agents. These approaches show that the more spatially distributed a system is, the more important it is to take care of local interactions and the differences between agents. In energy systems, the spatial component is essential, especially if taking into account distributed measures, which are applied locally, such as is the case in the smart grid.

McArthur et al. [2007a,b] identify that the stumbling block for multi-agent systems in the power engineering domain is mainly a lack of experience with these types of systems in the industrial context, as well as the lack of standards. This is one of the reasons for a rather slow introduction of agent-based approaches in the sector, in comparison to other disciplines such as biology or social sciences, where standards and the development of real systems is not that relevant.

Chappin and Dijkema [2007] elucidate the impact of CO_2 emission trading on European electricity production through a specifically created agent-based model. The model emulates the long-term evolution of European electricity production as a series of investment decisions by independent agents. The results underpin recommendations for CO_2-policy.

Quite a number of ABMs exist on the energy market. They mainly focus on the economic effects of energy systems. Agents usually represent actors in the market, which behave autonomously and have some kind of influence on the market. In [Sensfuss, 2007], ABMs dealing primarily with energy markets are reviewed. Furthermore, an agent-based model of the German electricity sector has been developed. For this purpose, electricity demand and generation, as well as the electricity market were modelled through agent-based approaches. This work focuses on the economic effects of the introduction of renewable energy sources such as the effect on spot market prices of electricity. Other examples of energy markets can be found in [Jacobo et al., 2008], which models the electricity market of the United Kingdom; in [Tellidou and Bakirtzis, 2009] and [Chan et al., 2010].

Agent-based approaches have also been made on technical design and electrical and control engineering. Notably, models dealing with the control of small autonomous grids, called mini- or microgrid control can be found. First approaches on applying MAS for control of mini- or microgrids date back to Tolbert et al. [2001]. Hatziargyriou et al. [2005] apply MAS to the centralised and decentralised control of microgrids. Dimeas and Hatziargyriou [2005] propose a MAS based framework for the control of microgrids. The main objective is to present a system capable of integrating several functionalities and to propose a general scheme for the control of microgrids. Dimeas and Hatziargyriou [2007] present the advantages of using agents for Virtual Power Plant (VPP) control. More specifically this paper, through examples and case studies, explains how local intelligence and the social ability of the agents could provide solutions for the optimal and effective control of a VPP.

Other agent based models for microgrid control can be found in [Jiang, 2006; Oyarzabal et al., 2007; Pipattanasomporn et al., 2009]

In relation to the smart grid and future information and communication on a trans-national interconnected power grid, Tranchita et al. [2010] propose different suitable modelling approaches, among them ABM, which could help acquire a better understanding of the newly identified risks of including ICT into the energy system.

The author has contributed to the development of several agent-based models for energy system analysis over the past few years. In [Evora et al., 2011] the Tafat framework for simulating complex energy systems is presented, which includes important software engineering techniques like Model Driven Engineering. Based on a metamodel, models including different behaviours can be automatically created. To model domestic demand, a disaggregated model of electricity consumers is created. A case study simulating the load curve of 1 000 households, composed of five different social groups, is discussed and compared with an aggregated curve. The model represents the load curve of a sample of households using a bottom-up approach and is a promising, powerful and high performance tool for energy system modelling.

In [González de Durana et al., 2009] an example of a simple Hybrid Renewable Energy Systems (HRES) is presented and an ABM for such systems, a microgrid, oriented to designing a decentralised supervisory control was developed. In [Kremers et al., 2009] and [Kremers et al., 2011], an agent-based, multi-scale wind generation model is presented. This model will be discussed further in the next chapters. Viejo and Kremers [2009] present an ABM of an integral island energy system, which represents the high voltage grid and can be used for decision making on regional planning scenarios. Modelling microgrids through an initial complex system approach, based on agents is presented in [Kremers et al., 2010], where some first simulation results from an integrated load flow algorithm were shown. In [González de Durana et al., 2010] the importance of interaction among different layers of the systems was identified, as well as the need for systemic models which are able to deal with different layers and scales. The models presented in these publications serve as a basis for this work.

4.2 Modelling Approach

Agent-based modelling tools are able to capture complex system behaviour such as wanted and unwanted emergent behaviour in these systems, internal and external events, communications within the system, etc. In particular, local effects of the single units comprising the system can be modelled and their effects can be analysed at the system level. On the other hand, technical models of the single agents are necessary. Therefore, dynamic systems have to be represented in the form of the physical and mathematical models of the devices.

Furthermore, the large amounts of interactions and feedback appearing in distributed energy systems make linear simulation models inappropriate for systemic simulations.

Therefore, an agent-based approach has been chosen. This approach is combined with other complex-system-relevant tools, such as system dynamics and discrete event simulation. The agent-based approach allows the inclusion of other modelling methods, for example modelling the reasoning of an agent. They can also be coupled either at a parallel level or embedded in a hierarchical structure.

The agent-based approach was chosen because it is able to represent autonomous entities, which interact with their environment. While retaining a broad definition of what an agent can be, we can imagine representing even very simple entities of the energy system through agents. So, for example, an electrical heater has a constant operating power, which can be switched on and off. It interacts with the environment through heat dissipation and power consumption. Its behaviour is defined simply by a constant consumption. Modelling the entities of the system in this way, also allows an encapsulated and hierarchically structured representation, as we suppose that an agent (as an extension of *objects* in object oriented programming, and inheriting their properties) can have one or more other agents nested inside itself. What is important to remark is that, even if the agents are modelled in a simple way, the modular structure of an agent based model allows for the improvement of the behaviour of the entities at any stage. This enables the addition of extra features, such as an intelligent control, to the previously mentioned heating system. The degree of intelligence is not limited, so an agent can include any kind of decision algorithm, from the most simple to the most complex. The advantage is that the modelling system provides the potential for this from the beginning. This is because it has been conceived with these features in mind, and to overcome issues of non-modular or difficultly extensible approaches.

Using this type of approach allows the modelling of energy systems from a complex systems point of view (as proposed in Section 2.4). It also means that the design of the model has to be carefully chosen.

A classical, top-down approach is not sufficient to model a complex system. The representation of complex phenomena like emergence, requires that the system should be based on individual rules which describe it in a bottom-up approach. These kinds of models enable emergence. It should be ensured that these phenomena emerge from the model, and are not explicitly programmed into it.

4.2.1 Challenges of the Electrical System

Some of the challenges of future energy systems that could be tackled using a complex system approach through simulation with agent-based models are:

Extrapolation of results Dealing with smart grid demonstrators opens new research questions which were not addressed some years ago. When testing new smart grid technologies on pilot demonstrators or individual experiments, large-scale penetration is not contemplated. Results, based on small amounts of data should be extrapolated to a larger scale, on at least two levels:

- Larger penetration scales of a given territory (higher density of installations)
- Larger geographical spread (extrapolation to a geographically larger territory)

Furthermore, the capacity of ABM representing different scales simultaneously can help to solve the uncertainties of changing the scale of applications.

Centralisation - decentralisation Using a bottom-up approach allows the modelling of decentralised decision making and, at the same time, the connectivity of agents enables a centralised control. Therefore, ABMs are suitable for the comparison of both centralised and decentralised control strategies on the same energy system.

Adaption and evolution in dynamical systems Models based on agents are easily modified on a unit basis, and therefore can truly reproduce adaptive and evolutionary mechanisms. Modelling based on agents focuses on individual units and therefore these units can be hybrid and don't have to be totally homogeneous, as it is the case in energy systems. Through their reactivity to the environment, these models are dynamic in an unstable equilibrium. This characteristic of ABMs can reproduce unexpected behaviour on the system, like snowball effects.

Representation of technical as well as social models The use of agents as encapsulating units in a general sense, allows the inclusion of many types of models from different fields. Through standardized interfaces it is possible to combine these models in the same framework. This, for example, permits an interaction between sociological and technical models.

Behavioural changes The reactivity of ABM and the possibility of evolution over time allow the implementation of models that take into account different behaviours on the same environment. Therefore they are suitable for models where behavioural changes have a strong impact.

4.2.2 Multi-Scale System Modelling

Multi-scale modelling is an emerging subject of research and development, in which systems are represented in different structural, spatial and temporal scales. These systems

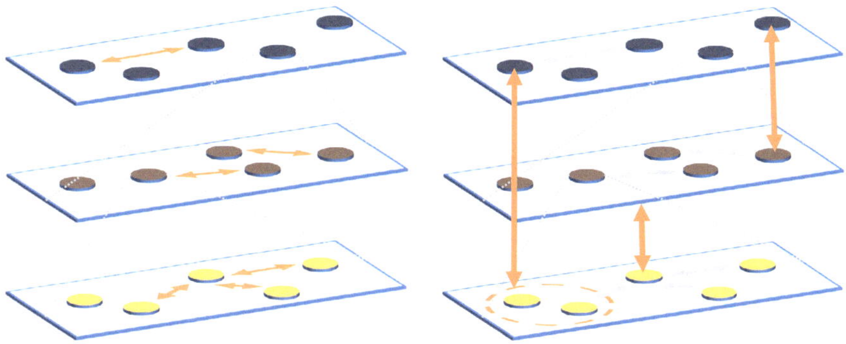

(a) Horizontal interactions (b) Vertical and horizontal interactions

Figure 4.2: A multi-scale system with different layers and horizontal (4.2a) and vertical (4.2b) interactions. The multi-scale system is composed of different layers; the top layer represents the whole system, whereas the layers below represent sub-systems at other scales. On the left, interactions are only present within a given scale. On the right, cross-scale interactions are added, which aim to reflect effects that encompass different scales. These are what we call cross-scale interactions.

show important features that are the objects of the modelling and should therefore be represented on these different scales. Multi-scale phenomena can involve complex processes, such as feedback loops between different scales.

The topology and structure of the energy system has different scales itself, which are not directly linked to spatial or temporal scales. The low voltage grid, the distribution grid and the transmission grid can all be seen as different structural scales within a complex system. They must be taken into consideration in this systemic representation.

Multi-scale simulation systems offer several advantages over classical models. The ability to run simulations on different scales using the same model is a relevant topic for the future modelling of energy systems. The need for this kind of simulation emerges from:

- Distributed generation that has to be integrated at almost all scales (and not only in the high voltage grid as was previously the case).
- Demand which is no longer considered as a *static* and uncontrollable mean. At lower levels of the system, local demand side management can be performed which can have an effect at higher scales.
- Monitoring that is implemented at lower scales of the electrical system (local measurements) will allow the analysis and optimisation of its operation at these lower scale levels.

Furthermore, simulating several scales in the same model has the advantage that there are fewer models and no need to port data between platforms. This leads to a more efficient simulation run and decision-making support. The challenges of these kinds of simulations are that a multi-scale model, at present, is not as accurate as a purpose-built model. So, the modelling method, the parameters, etc. included must be carefully chosen to ensure both flexibility and accuracy. Further attention has to be put on the inter-scale interactions. This is a complex issue, as multi-scale systems require both horizontal and vertical interactions. Horizontal interactions should be easily managed, as they take place within the same scale. Vertical, cross-scale interactions enable the representation of complex system phenomena. These cross-scale interactions must be modelled carefully to a design that that enables them. This is not an easy task.

Figure 4.2 shows these interactions in a multi-scale system. Current agent-based models usually take place only at a fixed scale and in general do not encompass different scales or layers. Adding multiple scales to the representation of the system increases the degree of complexity. However, only in this way can complex interactions at different levels of the system be represented.

Multi-scale models usually also have to deal with different spatial and time resolutions, which should be adapted to each scale. When modelling cross-scale interactions, different representations have to be considered. Geographical information systems (GIS) can be used to represent different spatial resolutions and the location of the entities,.

4.3 Socio-Technical System Models

Social sciences use simulation modelling as a tool in order to analyse and understand phenomena related to human behaviour. Behavioural models include a large degree of uncertainty, as their aim is to represent difficulty predictable, or only statistically characterisable, actions. These models have therefore to be non-deterministic in order to capture the heterogeneity of the actors. Behavioural models can only be validated with difficulty, due to their complex decision making processes.

On the other hand, technical models aim to represent the operation of technical systems. These systems (in the real world) are usually intended to be deterministic, or at least, their mode of operation is completely known as these systems are, in fact, man-made. Models representing those systems are directly aimed to recreate the system's mode of operation, for example based on its physical laws or discrete states. Their operation can be usually modelled in an unequivocal, predictable way and is deterministic.

Socio-technical systems are present in the real world everywhere in which a technical system interacts with human behaviour. This is the case for almost all technical systems. In order to model the "whole" behaviour, human behaviour has to be added to technical models. Coupling social with technical models is therefore a requirement for modelling

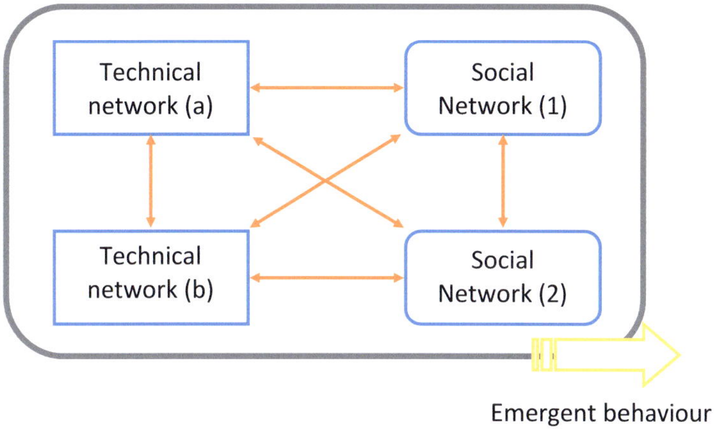

Figure 4.3: A socio-technical system with multiple networks and emergent behaviour at system level.

large, complex systems. Agents are well suited to this approach, as by definition, they include behaviour which can be interpreted as a social or also technical behaviour. Both systems have interactions with the environment, and they interact with each other, either directly or through the environment. However, it can be said that technical models are rather deterministic, and that the agent definition includes a proactive behaviour which is not always the case for technical systems. In any case, a simple technical behaviour can also be seen as a behaviour in itself. More important is its autonomy concerning its behaviour and the fact that a technical system always has the objective for which it was designed and operates with the intention of fulfilling this objective.

A socio-technical system consists of one or more social network and one or more physical or technological network that interact with each other. The different types of networks follow, on the one hand social and on the other hand, physical, laws. The socio-technical system itself is influenced by both types of laws [Van Dam, 2009].

Modelling socio-technical systems is, like the systems themselves, a complex task. Modelling very different aspects is required, and the modelling approaches in the social and the technical areas differ significantly. Different solutions can be found for this. For example, each original model can be kept in its actual approach and be coupled exogenously; or a more convergent approach can be taken, by combining the different approaches through multi-method modelling (Section 3.5. In the same work, Van Dam [2009] reviews some papers presenting approaches to modelling socio-technical systems.

4.4 Simulation and Modelling Tools used

The paradigm shift in the energy sector involves new challenges for the modelling and simulation of energy systems. More decentralised models that can reproduce the autonomous behaviour of the different units of the system are needed. The models currently used rely largely on experiences gained over the last decades, and are thus suited to hierarchically controlled one-way systems. A review of tools is presented in Valov and Heier [2006]. This includes several commercial tools used in power engineering, such as Eurostag or PSS/E and a number of non-proprietary tools, such as several toolboxes based on Matlab/Simulink, for example Matpower Zimmerman et al. [2011] or PSAT Milano [2009].

Matlab is probably the most widely used simulation program for control systems at the academic level, although there are many other programs like Maple, Mathematica, Octave, Scilab, etc. that are also very common. With Matlab, simulations are possible in a mathematical sense, i.e. to apply numerical methods to solve differential equations representing the system. However, as it is designed using technologies from the 1950's-60's, it lacks the advantages of more modern object-oriented software. These advantages are evident if the system to model is of discrete event type and even more so if it is a *hybrid* or an *agent-based system*.

Taking this approach, we will look at the problem from a new perspective, trying to offer new solutions with an innovative approach based on multi-method modelling. This work presents the opportunities and possibilities of a complex system approach to energy systems, as well as being an example of how modelling and simulation can help to examine and analyse potential smart grid architectures, configurations, and strategies prior to implementation.

Other agent-based tools are:

- Swarm
- Repast
- Netlogo

Anylogic is a well-regarded program in the community of multi-approach simulations, but little known in the areas of automation and control engineering. Anylogic is based on the latest advances in object-oriented modelling applied to complex systems [Borshchev and Filippov, 2004]. It currently supports three approaches or *modelling paradigms*:

- System Dynamics (SD)
- Discrete Events (DE)

- Agent Based Models (ABM)

These three approaches are mutually compatible, so that for example, in modelling a hybrid system we might use the SD method for the continuous part of the differential equations and the DE method for modelling the events. Anylogic models are multiplatform, portable Java applications that can run on their own anywhere a Java Runtime Environment (JRE) is installed. The models can be also run in a web browser as a Java Applet, which is an easy way of publishing the models. Moreover, Anylogic allows the development of animations of active objects: the assembly of the image is done automatically. In this way the animations are highly reusable and can be displayed on applets.

In Anylogic, scalable models can be easily created, because it is possible to define arrays of objects whose size is a parameter of the model. It is possible to even reflect the dynamic changes that can occur in a real system, by adding or deleting items, or changing their interconnection, during run time. Some of the simulation algorithms have been modified to work in hybrid environments (hybrid state machines) combining both continuous and discrete systems.

For continuous systems, we use the SD approach. This approach allows us to directly insert differential equations into the models, which are solved in Anylogic. As it uses the SD formalism there is a clear overview of the dependencies of stocks and flows. In Appendix D how to model differential equations in SD in Anylogic is shown.

Furthermore, Anylogic includes an optimization engine based on the OptQuest package from OptTek Systems. These tools allow a framework for simulation model runs which includes parameter variation experiments, calibration (through optimisation) of parameters, optimisation, and Monte-Carlo runs.

4.5 Environmental and Production Models

Production of electricity can be mainly classified as conventional or renewable, from the point of view of the primary energy source used. In the following section, we will address renewable energy sources (RES), as they are especially relevant for modern grids.

Some of the RES are strongly correlated to environmental conditions. This is particularly the case with wind and solar power; although other RES might also be affected by meteorological effects. In the following example, the production models shown are used, making a special focus on RES. This has allowed greater focus on the fluctuating RES, which introduce stochastic and unpredictable magnitudes into the energy system.

4.5.1 Wind Power Generation Model

This model[2] aims to represent wind power production by modelling wind farms consisting of wind turbine units on different time scales, ranging from short (minutes) to long-term simulations (months). The model takes into account fluctuating wind speeds and technical reliability. It can compute the aggregated output power of the wind farm influenced by different random factors and can, thus, recreate a realistic power unit for integral energy system simulations. The simulation of this data is performed in real time, so that power output at a specific time can be reproduced and injected into the energy system simulation.

Stochastic Wind Speed Simulation

Generating realistic wind speeds is an important task for analysing the effects of wind production in an electricity system. Fluctuating wind speed is the origin of the temporal variation of power injected by this production type and thus has direct effects on the production-demand balance and grid stability. One of the challenges of wind speed simulators is to reproduce the different scale term fluctuations, as described in [Nichita et al., 2002]. To this end, different models have been developed. The model chosen here is built up in two steps, comprising slow and fast components, [Bayem et al., 2008] with some minor modifications. More accurate wind models (that take into consideration e.g. long-term [Billinton et al., 1996] or cross-correlations [Allerton, 2008]) are available, but this one is sufficient for the purposes of this work. An overview of some more approaches can be found in [Aksoy et al., 2004]. It is important to add that to get a realistic simulation of a specific site, records of historical data are needed to obtain the parameters of the model, as even the best model is useless if not accurately fitted.

The slow component The first part, which is used in previous works of the author [Kremers et al., 2009; Viejo and Kremers, 2009] is a generator of hourly mean wind speeds. This time series model is based on an ARMA (Auto-Regressive Moving-Average model which is given by

$$\begin{aligned} y_t &= \phi_1 \, y_{t-1} + \phi_2 \, y_{t-2} + \ldots + \phi_n \, y_{t-n} \\ &\quad + \alpha_t + \theta_1 \, \alpha_{t-1} + \theta_2 \, \alpha_{t-2} + \ldots + \theta_m \, \alpha_{t-m} \end{aligned} \tag{4.1}$$

The data series y_t is used to build the model, i.e. to calculate the auto-regressive $\phi_i; \ i = 1, 2, \ldots, n$ and the moving average parameters $\theta_j; \ j = 1, 2, \ldots, m$. $\{\alpha_t\}$ is a Gaussian white noise process with zero mean and standard deviation of σ_α which is part of the moving average (MA) part of the model. Considering the orders, the process is referred to as ARMA(n, m). The parameters used in this work were chosen from an ARMA(3,2) approach, but the model was developed up to ARMA(4,3) and can be easily adapted to

[2] The following sections are inspired on the works published by the author in [Kremers et al., 2009] and the book chapter [Kremers et al., 2011].

other orders. For example, a pure AR(2) model [Aksoy et al., 2004] which was previously implemented can be seen as an ARMA model with $n = 2$ and $m = 0$. The order of the model depends on the quantity of historical data available, since, if there is only a little data, an accurate model cannot be reached even with higher orders. There is a range of literature available on parameter estimation. Fitting models are normally based on the least squares regression methods that try to minimise the error value. For AR parameter estimation, the Yule-Walker equations are widely used.

The simulated hourly mean wind speed [Billinton et al., 1996] can be obtained by

$$\overline{v}_1(t) = \mu + y_t \tag{4.2}$$

where μ is the mean wind speed of all the observed data. If observed hourly mean speeds μ_h and standard deviations σ_h are available, a more realistic simulated wind speed[3] can be calculated as:

$$\overline{v}_2(t) = \mu_h + \sigma_h \cdot y_t \tag{4.3}$$

The fast component A detailed model was needed which could not only compute hourly mean wind speeds because temporal scalability is necessary. The ability to reproduce realistic wind speeds in real time can be achieved by adding a so-called fast component to the previously described, slowly varying signal. For this purpose turbulent phenomena are modelled by a highly fluctuating signal given in [Bayem et al., 2008] by the following differential equation:

$$\frac{dw}{dt}(t) = -\frac{w(t)}{T} + \kappa v_h(t)\sqrt{\frac{2}{T}}\xi(t) \tag{4.4}$$

where $T = L/\overline{v}$, being L the turbulence length scale, κ a factor that depends on the geographical location of the wind turbine site [Welfonder et al., 1997], $\xi(t)$ a Gaussian white noise and $v_h(t)$ the hourly mean wind speed. The equation describes a stationary Gaussian process. This component allows us to generate a time continuous signal that represents a real time wind speed.

Wind Power Generation Modelling (Turbines)

Turbine model There are plenty of technical models for wind turbines. The model used here is a generic approach, which takes into consideration the agent-based approach of the framework. As the wind turbine has to be able to be replicated (in order to create wind farms with tens of turbines, or more) a simple model was chosen to ensure fluid simulations. The basis of this model is the relationship between the power output of the turbine, which is a function of wind speed actuating on its rotor blades. Three different models that are commonly used have been identified in the course of this work. The *real model* is not a mathematical model itself. It just shows the $P(v)$ curve of a specific turbine

[3] The method is explained in detail in [Billinton et al., 1996].

- based on the manufacturer's data. In general, the curve has a shape similar to the one shown in Figure 4.4a.

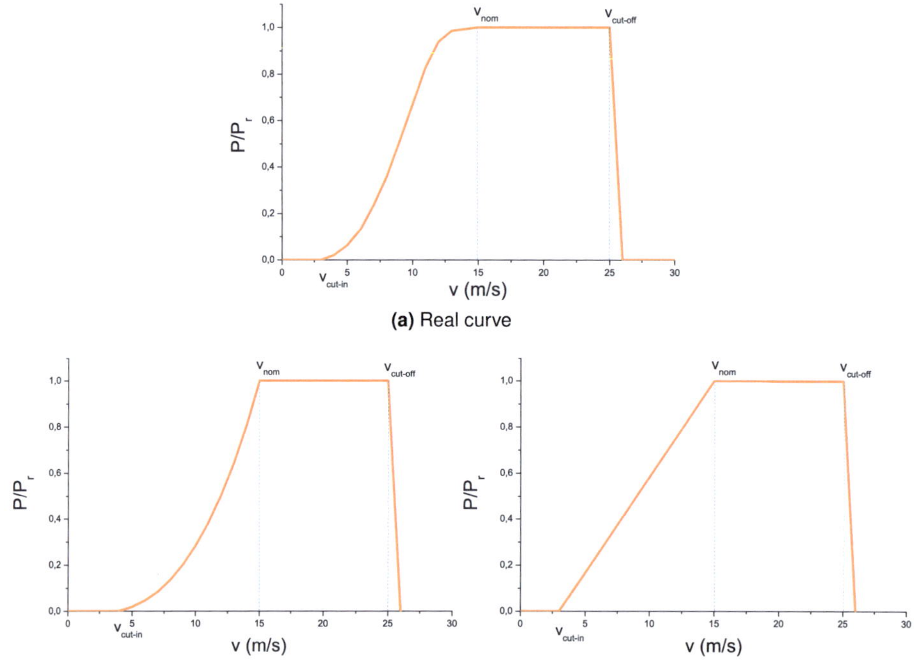

(a) Real curve

(b) Polynomial approximation

(c) Linear approximation

Figure 4.4: Sample wind power curve and two approximated curves. P_r is the rated power.

The curve shows the typical profile of a wind turbine. The cut-in speed is the minimum wind speed at which the turbine can start to work, the nominal wind speed is the point at which the rated power of the turbine is achieved. This power is normally almost constant until the cut off wind speed is reached, at this point the turbine must be shut down to avoid damage caused by too strong winds. So, four principal working states can be defined as:

- Stopped: for $v < v_{cut-in}$
- Partial load: for $v_{cut-in} < v < v_{nom}$
- Rated load: for $v_{nom} < v < v_{cut-off}$
- Cut-off: for $v > v_{cut-off}$

The transitions between these states are smooth because of the technical characteristics of the rotor and generator in the real curve. The most interesting state to be observed is the partially loaded state, where the turbine shows a non-linear $P(v)$ dependence. Here the start dynamics of the turbine, as well as the adaptation to the fully loaded capacity at rated speed, can be observed. This phase can be approximated by a polynomial term as shown in Figure 4.4b. The polynomial model assures the curved shape of the curve, but the trace just before achieving the nominal wind speed is idealised. The linear approximation of the curve, which is used in more simplified models, can be defined by a linear interpolation of the values for v_{cut-in} and v_{nom}. It can be seen in Figure 4.4c. The last model might be used only when the characteristic wind speeds of the turbine (and no power curve) are available. Though, the polynomial approach can be also be used as an approximation by using a polynomial of degree three as described in [Chedid et al., 1998].

The cut-off state is reached when the turbine gets shut down because of exceeding $v_{cut-off}$. Further, a $v_{cut-back-in}$ parameter can be defined for the model. Its value denotes the wind speed at which the turbine restarts after being in the cut-off state. This value describes the restart process of the machines after strong wind periods.

Being MTBF the Mean Time Between Failures of a unit defined by

$$MTBF = \frac{1}{\lambda} = \frac{operational\ time}{number\ of\ failures} \tag{4.5}$$

where λ is the failure rate. Using $MTBF$ allows modelling the availability of a wind turbine over time. The equation describing the Mean Time To Recover

$$MTTR = \frac{down\ time}{number\ of\ failures} \tag{4.6}$$

is also included, where $down\ time$ is the time when the turbine is inactive because of a failure, maintenance or reparations. The MTTR is therefore an indicator for the average down-time between an incident and the following start up. A failure model is integrated into the turbine model which takes into account these two parameters. The rates (inverse values of them) are used to determine failure probability used in the transition between states.

4.6 Consumption Models

As we have seen in previous chapters (Section 2.3.2, there are different methods for modelling electrical consumption. As discussed, there is a need for high-resolution models in order to clearly understand the composition of a load curve. Aggregated models, like standard load profiles, can approximate the load curve for a large number of consumers, but might be wrong for a particular population or region. In order to have a high-resolution model of domestic demand, several models of household appliances were developed [Costa Gómez, 2011]. Called *technical models* they represent the device's consumption load curve. However, in many cases this is not enough to represent the consumption of a household, as these technical models do not include the associated human behaviour. They are therefore, like an unused appliance. Additionally, we need a social behaviour agent which controls the devices. This agent is the most complex part of the model. The social behaviour agent will decide when to switch on or off certain devices, as a human being would do. However, the more device there are in a household, and the more persons living within, the more complex the different dependencies and correlations of the usage of the devices will become. Initial approaches to this were made in [Evora et al., 2011] in which 1000 households were simulated by using a simplified model of social behaviour.

As we can see, the micro-modelling of domestic demand is a quite complex task. In the given context, we will address one appliance, a refrigerator, in order to be able to illustrate a concrete case study where the impact of the given appliance can be clearly observed. A simple, social behaviour is included to represent the use of the refrigerator; it is conditioned by door openings followed by the insertion and extraction of items.

4.6.1 Modelling a Refrigerator

A refrigerator was chosen as an example of consumption device. Refrigerators offer a more complex behaviour than static loads that do not vary over time (like lighting or entertainment devices), but are continuously connected to the energy supply. Therefore, refrigerators offer an interesting potential for demand side management. Acting as thermal storage, a refrigerator is capable of being disconnected for some time, and then can later recover the lost energy, without having a high impact on the quality of the service it offers. In order to study and analyse a refrigerator's potential for load shedding, a detailed model

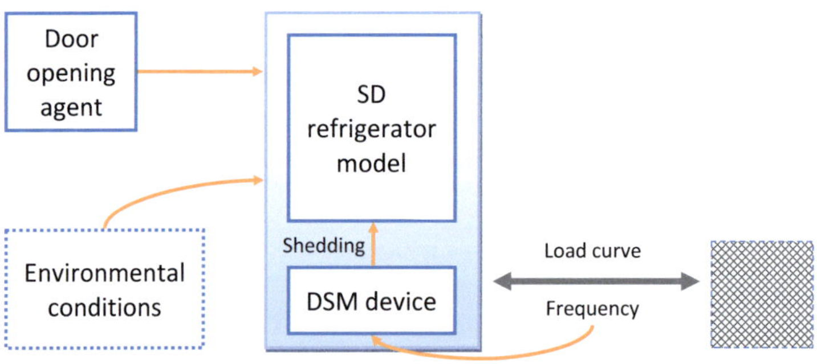

Figure 4.5: Overview of the refrigerator model: the core thermal model (called SD refrigerator model) is coupled with a DSM device which permits load shedding, and additionally with an environment and a door opening agent. Finally, the electrical part of the model is coupled with the energy system.

is described which will later be implemented in simulation. The structure of the model is based on a thermal model that describes the thermal processes in the refrigerator (called SD refrigerator model). It is coupled with a DSM device to unplug the refrigerator under certain conditions. This performs a load shedding, which allows the refrigerator to be disconnected if there is a need for load reduction. Moreover, the refrigerator model is coupled to the environment (mainly the environmental temperature) and a door opening agent. The door opening agent simulates manipulations of the refrigerator, as it has been identified that door opening has a major impact on the internal temperature, and thus impact the compressor's working cycles. The model is obviously also linked to the energy system, by its generated electrical load. Additionally, the sensors of the DSM device can capture data that gives information about the state of the grid. We see here a clear agent-based structure (see Figure 4.5), which describes capturing values from an environment, taking decisions (for example in the DSM device) and acting on the agents (door opening, DSM disconnections) which can affect the energy system. The agents are clearly separated and well-defined, as are the interactions between them.

First, we will describe the core model of the system, which is the refrigerator itself. A refrigerator is a device based on a vapour compression cooling system whose main components are the condenser, the evaporator, the compressor and the expansion device. See Figure 4.6.

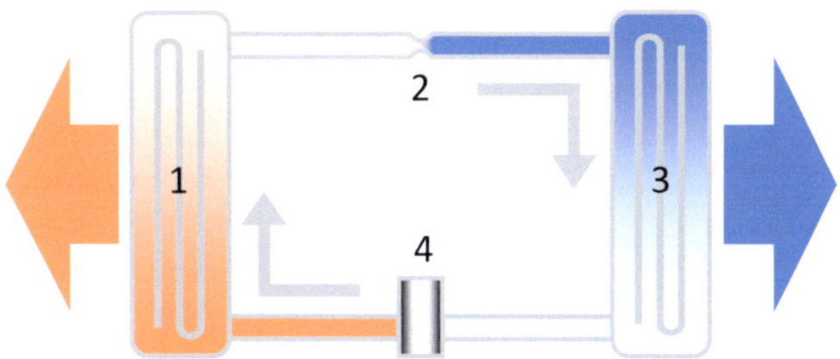

Figure 4.6: Operating principles of a refrigerator with its components: (1) evaporator, (2) expansion valve, (3) condenser and (4) compressor

- Compressor: attached to the electrical grid; it increases the pressure of the gaseous refrigerant.
- Condenser: a heat exchanger that dissipates the heat absorbed in the evaporator and the compressor energy. In the condenser the refrigerant changes its phase from liquid to gaseous state.
- Expansion valve: the liquid refrigerant flows though the expansion valve where its pressure and its temperature are reduced.
- Evaporator: the refrigerant at low temperature and pressure passes through the evaporator. Like the condenser, the evaporator is a heat exchanger and absorbs heat from

its environment. The liquid refrigerant, which flows into the evaporator, is transformed into gas by absorbing heat from inside.

The refrigerator model used in this study is based on the heat fluxes of its internal cell, the evaporator and the exterior. These fluxes are described by a set of differential equations which characterise the heat transfer processes. The principles of the models are describe in [Cerri et al., 2003]. This model was extended to include door opening and charge effects. Door opening has a significant impact on the internal air temperature, especially when the refrigerator is situated in a warm environment. The food and drinks in the refrigerator also are relevant for the dynamics and inertia of the inner cell temperature, and function as heat storage. The evolution of internal temperature is described by a differential equation that takes into account the heat transfer process between the external environment and the refrigerator cell and the cooling contributions by the evaporator plate, as well as direct air exchanges due to door opening and heat exchanges with the contents:

$$\frac{dT_i}{dt} = \frac{T_e - T_i}{\tau_{e-i}} + \frac{T_p - T_i}{\tau_{p-i}} + \frac{T_l - T_i}{\tau_{l-i}} + Q_{door} \tag{4.7}$$

where T_i is the internal cell temperature, T_e the temperature of the environment, T_p the temperature of the evaporator plate. τ_{e-i} is a time constant describing the heat transfer between the internal cell temperature and the environment. This constant depends on the thermal capacity of the cell, the surface of the walls in contact with the exterior and the heat transfer coefficient of the insulation of the cell. τ_{p-i} is the time constant for the heat transfer process between the evaporator and the inner cell. This constant depends on the thermal capacity of the inner volume, the surface of the evaporator and the heat transfer coefficient between the evaporator and the inner cell. T_l is the average temperature of the assumed content in the refrigerator. τ_{l-i} describes the heat transfer time constant between the charge and the internal air temperature, and will depend on the surface and type of content. Q_{door} describes the effect of a door opening.

The variation of the temperature of the evaporator plate is obtained by

$$\frac{dT_p}{dt} = \frac{T_i - T_p}{\tau_{i-p}} - \frac{\delta_c \cdot Q_f + Q_{ext}}{C_p} \tag{4.8}$$

where τ_{i-p} is a time constant describing the heat fluxes from the plate to the inner cell, depending on the thermal capacity of the plate, its surface and the heat transfer coefficient between the plate and the inner cell. Q_f is the cooling capacity of the plate, which is multiplied by a switching variable δ_c to denote that this capacity only affects the system when the compressor is working. The value is $\delta_c = 1$ for the working phase and $\delta_c = 0$ when the compressor is idle. C_p is the thermal heat capacity of the evaporator fluid and Q_{ext} an exponential function of time which describes the effect of released thermal power after the compressor switches off and is explained in detail in [Cerri et al., 2003].

The thermostat controlling the compressor turns it on when the temperature drops above $T_{p,max}$, but doesn't turn it off again until the temperature falls below $T_{p,min}$. Hence, the value of δ_c is triggered by two temperature thresholds in a hysteresis process, where:

$$\delta_c = \begin{cases} 0 & \text{if } T_p < T_{p,min} \\ 1 & \text{if } T_p > T_{p,max} \quad \text{until } T_p < T_{p,min}. \end{cases} \tag{4.9}$$

When the compressor switches off after a refrigeration cycle, the thermal power is released with the cooling fluid, passing from the condenser to the evaporator plate through the expansion valve. This phenomena is described in Cerri et al. [2003] as an exponential function of time.

The opening of the refrigerator door creates an immediate heat exchange between the environment air temperature and the interior, usually by warmer air flowing into the cell. This heat exchange is added to Equation 4.7 as an additional term:

$$Q_{door} = \frac{\delta_d \cdot (T_e - T_i)}{\tau_d} \tag{4.10}$$

The term is characterized by δ_d which expresses the heat exchange due to door openings. $\delta_d = 1$ only if the door is open; if it is closed it equals zero. The time constant τ_d represents the heat transfer by air exchange between the environment and the inner cell.

The effect of the load contained in the cell can be described as a heat transfer process between the internal cell temperature and the temperature of the contents.

$$\frac{dT_l}{dt} = \frac{T_i - T_l}{\tau_{i-l}} \tag{4.11}$$

where τ_{i-l} is a time constant depending on the type of contents (food, drinks, etc.), its thermal capacity and its surface in contact with the air of the inner cell.

Finally, the electrical power P_{el} is described by the coefficient of performance and the cooling capacity of the plate:

$$P_{el} = \frac{Q_f}{COP} \tag{4.12}$$

4.6.2 Inclusion of Social Behaviour through Door Opening

In addition to the technical behaviour of the refrigerator, the operation of the compressor cycles is strongly effected by the refrigerator's use. Door opening and (warm) contents placed into it can prolong refrigeration cycles and thus prolong consumption. In order to recreate and include these uncertainties, a stochastic model has been included to represent the human behaviour that causes door opening. This door opening agent controls the refrigerator door and is embedded in the refrigerator. The model is based on a discrete event state-chart, which triggers the δ_d values of the differential Equation (4.10).

The agent implements a random door opening, based on an average daily rate. The average number of daily door opening n is defined by μ_n. The agent opens the door randomly, according to an exponential distribution, with a mean according to the given parameter. So for example, the mean of random openings per day will have a certain value, which can be higher or lower in a particular case, as this is a non-deterministic model. In any, case, over a longer period or several refrigerators, the average will tend to $\overline{n} \to \mu_{openingsPerDay}$.

The second set of parameters defines the duration of each opening. This is calculated by a random number, which is taken from a normal distribution with the mean defined by the mean time $\mu_{doDuration}$ and the standard deviation $\sigma_{doDuration}$ of the door opening.

4.7 Storage and Smart Grid measures

Electrical storage plays an important role in smart grid technologies. Being able to store energy and release it at another time allows a more flexible use, and control, of resources.

4.7.1 Generic Storage Model

The term *storage systems* is intentionally used to describe the variety of implementing storage in an energy network. The different technologies within which the systems are realised (batteries, compressed air, etc.) are not relevant in this model. An abstract model of a storage system will be used to make a generic approach to the problem, rather than attempting to solve it using developing technologies with all their constraints .

The main model is described by the following equation,

$$\frac{dE(t)}{dt} = P_{in}(t) - P_{out}(t) - P_{loss} \tag{4.13}$$

where E is the energy content of the storage, P_{in} the effective in-flowing power, P_{out} the effective out-flowing power, and P_{loss} the power lost in storage. These describe the losses over time when storage is in the static mode (no in- or out-flows), but in which the amount of energy gets reduced over time. Usually they are given in units like kWh per month, for instance. It can be a function of E, in a detailed model. Conversion losses are added by an efficiency η_{in}

$$P_{in} = \eta_{in} P'_{in} \tag{4.14}$$

which describes the relationship between the real in-flow power P'_{in} and the effective in-flow. Analogously this can is defined for the real output power P'_{out}

$$P'_{out} = \eta_{out} P_{out}. \tag{4.15}$$

The model can then be parameterised with typical values (see Table B.1 for a given technology and in this way be adapted to represent it.

4.7.2 Frequency-Based Load Control

Alternating current (AC), which is mostly used in current power systems is transmitted at a certain frequency. The frequency is the number of occurrences of electric charge direction reversing per unit time.

The balance between electrical supply and demand must be managed, especially for a fragile grid. During a day, production is constantly adapting to needs, starting or stopping production units as and when necessary. Frequency based load shedding is common in island grids because of their lack of interconnections to neighbouring networks. This procedure aims to prevent a widespread incident, such as cascading events, which can be quite dangerous for the system.

A highly reactive demand response is needed to support the primary reserve. As soon as an abnormal frequency drift is detected, the demand side can react by adding or reducing power consumption. This method is also called under-frequency load shedding (UFLS). In contrast to classical reserve mechanisms, the demand side is made up of a large number of small consumers, rather than a big dispatchable power plant. However, if enough consumers are taken into account on the demand side and if they are properly managed, they can act as a virtual power plant (VPP), and thus have a mutual, aggregated effect on the system.

In the case study presented, refrigerators were chosen as loads, they are an appliance with a very high penetration rate, and even without high peak powers, refrigerators are usually continuously plugged to the grid. Refrigerators have the potential to operate as a thermal storage medium, and can offer a flexible response to the demands of the energy system.

Different approaches on demand side management concerning refrigerators have been proposed although few simulations can be found. Notably Short et al. [2007] uses a simplified refrigerator model to simulate many of them. He proposes a dynamic operation strategy by varying the thermostat threshold linearly to the grid frequency. However, in any of these approaches, the effect of the different demand side management strategies has been analysed at system level and with a large penetration through a detailed model. Only a few high resolution studies and models exist on domestic demand [Wright and Firth, 2007; Widen et al., 2009], and they are not coupled to an energy system model.

4.8 Grid Models

4.8.1 Load Flow

To design an electrical network, engineers need to be able to estimate the voltages and currents at any place within the circuit. Usually, an undirected graph shows the grid topology, and mesh or nodal analysis is often used to set up an equation system and solve it for the voltage and current at any place in the circuit. Standard linear algebra methods are suitable for linear systems, but other nonlinear circuits can only be solved using estimation techniques or specialised software programs. The most common method used here is a load flow analysis.

In an electrical network with n nodes, each one is associated with a pair of complex numbers, V (voltage) and $S = P + Qi$ (apparent power).

The given variables are:

- $|V|$ and $\angle V = 0$ are given at the *reference node*
- $|V|$ and P are given at *control nodes*
- P and Q are given at *load nodes* (the other nodes)

Applying the electrical laws results in a nonlinear system of $2n$ equations with $2n$ real unknowns,

$$g(x) = 0, \tag{4.16}$$

where $x, g(x) \in \mathbb{R}^{2n}$, with the unknowns:

- P and Q for the reference or slack bus
- $\angle V$ and Q for control buses
- $|V|$ and $\angle V$ for load buses

To solve the system, the most common approach in the literature is to apply the Gauss-Jordan (iterative, of slow convergence) or the Newton-Raphson method, which is faster, but requires that the Jacobian matrix is calculated and inverted.

When talking about load flows [Bergen and Vittal, 2000], it is often assumed that the electrical network is static. In fact, it is assumed that all data values are known constants and the effect of actual load variations over time is studied through considering a number of different cases for which steady-state conditions are assumed.

Still, to analyse transient effects after a value change in some nodes (e.g. in case of connection and disconnection of loads or sources), the values of the unknowns have to be updated quickly. It is these values that are used for the control of the network or for its simulation. The implementation of the algorithm in Anylogic [Kremers et al., 2010] was successfully validated against a load flow calculation performed by PSAT [Milano, 2009].

4.8.2 Grid Frequency Model

The System Frequency Response (SFR) model [Anderson and Mirheydar, 1990] is a simplified model which describes the behaviour of an interconnected system when dealing with a large disturbance. This model omits many details and ignores small time constants, providing a model that is useful in approximating the system frequency performance and has been used by different authors in the field [Kottick et al., 1993; Lee Hau Aik, 2006]. Despite of the model's simplicity, the comparison with other system disturbances and detailed simulations, as shown by Anderson and Mirheydar [1990] are rather encouraging.

The model is represented in 4.7 in a block diagram in the Laplace domain, as is usual in control theory. It is composed of two main blocks in a typical feedback loop configuration, where the lower block balances the system. The upper block describes the initial response $\delta\omega$ of the system to a disturbance in the production-demand balance, P_d, taking into account the inertia and damping factor of the system. The lower block feeds a negative power P_m at the input of the system representing the simplified response of the production system (primary reserve).

The SFR model allows further understanding of how the system parameters affect the frequency response. In a detailed model of the electrical system, this is more difficult because there, frequency response is a very complex function of many system variables.

The effect of a sudden load disturbance on grid frequency can be summarised in two points by the given model [Anderson and Mirheydar, 1990; Terzija, 2006]:

- The initial slope of frequency change is proportional to the power imbalance, and also depends on the inertia of the power system.
- The values of the minimum frequency, and the new steady-state frequency, reached during the following transient process are proportional to the power imbalance and depend on the dynamic properties of turbines, governors, loads, and other system control devices.

The parameters of the SFR models are:

- H: inertial constant [s]
- D: damping factor [1]
- K_m: mechanical power gain factor [1]

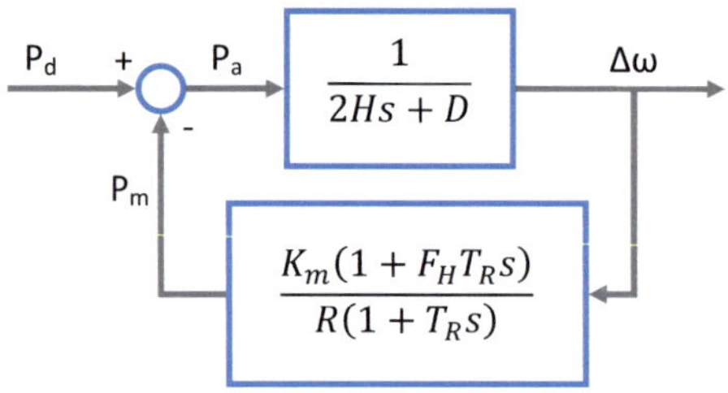

Figure 4.7: Simplified system frequency response model

- F_H: fraction of total power generated by the HP turbine [1]
- T_R: reheat time constant [s]
- R: governor regulation [1]

The inertial storage capacity of a power system is measured by the inertial constant, H, which is the number of full-output seconds of energy stored (assuming nominal frequency) typically varies between 2 and 8 s [Short et al., 2007].

4.9 Towards Electrical Energy models as a Whole

We have seen that is possible to represent many of the entities of an electrical energy system through simplified, technical and non-technical models. Production was exemplified by wind power, a renewable energy source. In this case, we saw the need for an environmental model representing wind speeds, because a purely technical model of a turbine is not enough to simulate wind power. Demand was exemplified by a domestic refrigerator. This appliance was chosen because of its high penetration rate and its continuous operation (continuously connected to the power grid). This device can be managed by a DSM. A simple model for under frequency based load shedding was created and a simple non-technical model represented the human behaviour of door opening was also used. The infrastructure, or network, can be represented by different models such as load flow or a simplified frequency response model.

These different modules, such as a wind turbine or a refrigerator, have been represented individually, following the agent-based philosophy. They were conceived to interact with each other: for example, the obtained wind speed can be used as an input for a wind turbine, or the generated load of a refrigerator is used in the grid model. Now, as a next step, the different modules can be *plugged* together, in order to create electrical energy system models, and run in simulation, in an interconnected way, at the same time.

Chapter 5

Integral Multi-Scale Case Study Models

After having identified and explored some of the models that describe different components of an energy system, an implementation in a modelling engine is performed in order to run simulations. Two case studies show the application of the models in a complex system environment. The first one focuses on production and the second example shows a demand model, which includes Demand Side Management (DSM) coupled to an energy system model. Both models show behaviours at different scales, on both the individual scale and on the systemic scale. The systemic scale shows a coupled model at a higher scale, interacting with the individual agents from the first model. Their implementation is based on the components that were developed in the previous chapter.

5.1 Wind Farm Case Study

The first case study addresses one of the most commonly used forms of renewable energy. Wind energy allows for efficient generation, through turbines located close to the rotor which directly convert mechanical into electrical energy. This case study presents an agent-based model of a wind farm, in which each turbine is modelled individually. Furthermore, failure behaviour is included to make the model more realistic. The model is coupled to a multi-timescale wind simulator, that allows for representing wind speeds with different accuracy.

5.1.1 Wind Simulator Implementation

Different modules were developed in Anylogic to build the wind simulator. Each module was encapsulated to work independently and has well-defined interfaces. This allows for different releases of the same module, which can be easily replaced.

The wind simulator modules as follows:

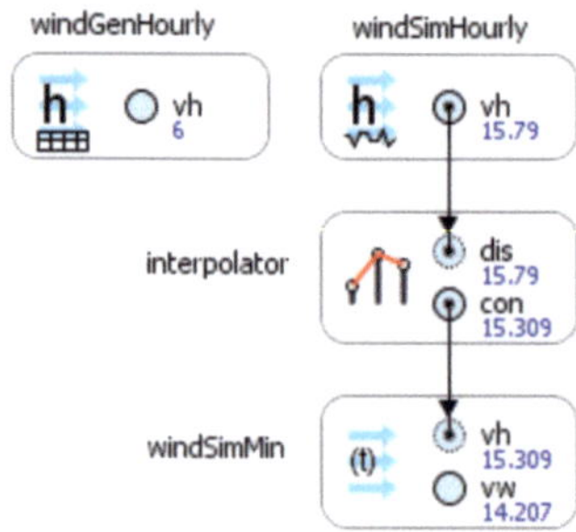

Figure 5.1: Modules of the wind simulator

- **Hourly speed module:** The hourly speed module provides the hourly wind speeds. In the current model, there are two possible implementations:

  1. The hourly wind speed generator is a module that uses a given dataset for speed generation. Normally it uses historical input, which gives hourly mean wind speeds. It can also be used to test extreme situations by simulating extreme conditions. Additionally, it allows for replicable simulation runs, by using the same time series as input for multiple simulations.
  2. The hourly simulator implements the slow component ARMA model described in Section 4.5.1. The parameters of the model are the hourly mean wind speed μ_h, the hourly standard deviation σ_h, the standard deviation σ_α of the $\{\alpha_t\}$ process and the AR and MA coefficients $\phi_1 \dots \phi_4$ and $\theta_1 \dots \theta_4$, respectively. The output generated is the hourly mean wind speed $v_h(t) = \overline{v}_2(t)$ by implementing the method described in Equation (4.3).

- **Detailed module:** The detailed module is needed for short time-scale wind simulations. The present release is a simulator. It is the implementation of the fast component and uses an average hourly wind speed as input. The input signal $v_h(t)$ is superposed with some turbulence. This can be fitted to real turbulence data by the parameters κ and L described in Section 4.5.1. The solution to the differential equation is computed by Anylogic's engine using the Euler method.

- **Interpolator module:** The interpolator is necessary to generate smoothed final wind speeds. As the hourly mean wind speed is calculated or given in discrete values for each step, the change of the mean would cause a non-continuous piecewise function with abrupt jumps in the final wind speed signal. Thus, a linear interpolation for the hourly wind speed was implemented. The module includes a parameter to determine the interpolation interval t_i measured in time steps of the current model time. There

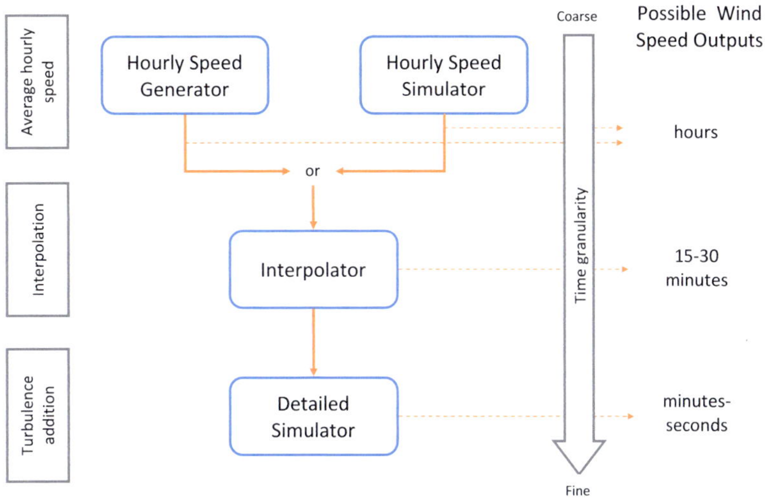

Figure 5.2: Time granularity of the model

is a connection between the hourly simulator and the detailed simulator, as shown in Figure 5.1.

The interoperability of the modules allows several combinations. For example, when historical data of hourly mean wind speeds are available and continuous values are needed, the wind speed generator and the detailed module can be used. However, if only statistical data on the site are given, the hourly wind speeds can be simulated through the hourly simulator based upon that data.

5.1.2 Turbine Implementation

The wind turbine is the core of wind power production. The requirements of the turbine were to convert wind speed to a suitable magnitude for the power system, i.e. injected power. This reflects the process of the wind turbine converting the wind's kinetic energy into electric energy through the generator. The wind turbine is modelled as an agent, because it will be replicated several times to create wind farms and each entity has similar, but not identical, characteristics. The agent can be customised through its parameters, which are shown in Table 5.1.

Making use of Anylogic's features to create hybrid models [Borshchev and Filippov, 2004], the turbine was modelled using the power curve model of the $P(v)$ relationship described in Section 4.5.1 in combination with UML state-charts. The power curve model was chosen to ensure flexibility in the application of the model. It is assumed that when modelling a wind farm, detailed information is available on the turbines used. This way, it is possible to customise each turbine with its corresponding power curve. The model of the wind turbine agent remains the same in any case.

The state-chart elaborated here is classified in states dependent on the output power and failure state. The three working states of the turbine are as follows:

- OFF: this state is active when the turbine is not producing any output power, regardless of the cause (no wind, overly strong wind speeds, etc.) except in the case of a failure
- FAILURE: this state is achieved when there is a failure or shutdown of the turbine due to maintenance.
- ON: the turbine is in this state when producing output power, regardless if the rated power is gained or the turbine is only partially loaded.

The transition conditions between the states are defined by the wind speed for the transitions between the ON and OFF states, and by the corresponding rates of the MTBF and MTTR in the case of transitions to and from the FAILURE state, respectively. The MTBF is used for both transitions from the ON and OFF states. The rates are always adapted to the current timescale by a factor that is proportional to it and set automatically by the model as a function of the scale chosen.

For the computation of output power, Anylogic's so-called action chart is used to link both the discrete state-chart approach and the continuous power curve. The output power is only taken from the power curve if the current state is set to ON. The state-chart and the action chart are shown in Figure 5.3.

Parameter	Description	Value
P_{nom}	Nominal power	275 kW
v_{cut-in}	Cut-in wind speed	3 m/s
$v_{cut-off}$	Cut-off wind speed	20 m/s
$v_{cut-back-in}$	Cut-back-in wind speed	18 m/s
$MTBF$	Mean Time Between Failures	1900 h
$MTTR$	Mean Time To Recover	80 h

Table 5.1: Wind turbine parameters

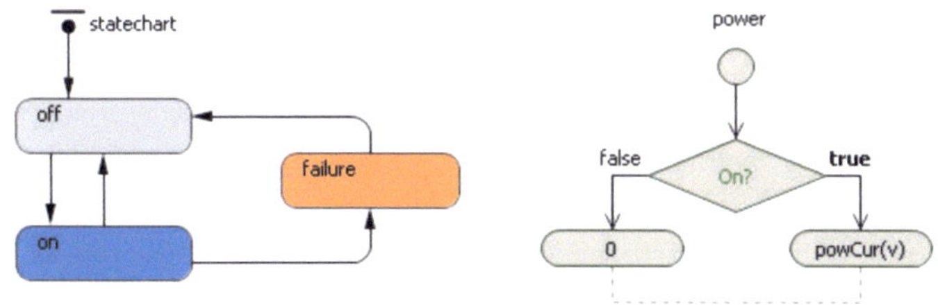

Figure 5.3: State-chart (left) and action-chart (right) of the wind turbine model including failure behaviour.

5.1.3 Implementation of a Wind Farm

After implementing the basic elements of our simulation, the wind turbine agents are grouped into an environment that defines common values for all agents within it and creates a framework among them that allows us to extract common statistical data. For instance, the aggregated output power of the wind farm, or the mean power by turbine could be computed.

A wind farm with 25 wind turbines is generated in the current example, as this is a typical number for medium size onshore wind farms. The power curve of the generators is the same for all, since it is assumed that the same types of turbines are installed. The power curve used here is inspired by the turbine type GEV MP 275, manufacturered by Vergnet Eolien. It has a 32 m diameter rotor and a rated power of 275 kW and is especially designed to be used in remote locations and can sustain hurricane winds when secured to the ground.

The wind parameters for the wind simulator were taken from models developed previously. The ARMA coefficients and the parameters L and κ were taken from real sites and from literature, described in [Kremers et al., 2011].

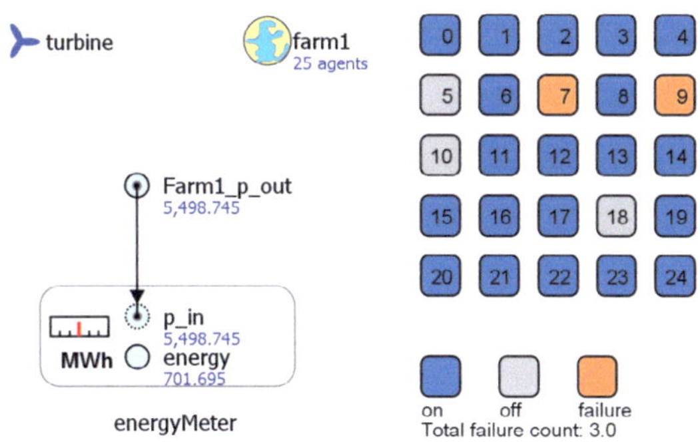

Figure 5.4: Representation of the states of the turbines composing a wind farm

5.2 Frequency-Controlled Refrigerator Case Study

For creating a systemic model of a refrigerator population within an energy system, we need to represent both the individual refrigerators that comprise a population, and an energy system which allows us to represent the behaviour of the grid. Both parts of the model need to interact, as the load curve generated by the refrigerator affects the energy system, and, conversely, a frequency based load shedding on the refrigerators requires the grid frequency values. Therefore, we create a cross-scale interacting model, where individual agents (representing each a refrigerator) at the low voltage level interact with a higher level frequency response model, in a bi-directional way. First, the refrigerator implementation is shown, followed by the frequency response model. Then the integration of both in a systemic model is discussed.

5.2.1 Implementation of the Refrigerator Model

The refrigerator model described in the previous chapter is based mainly on a differential equation, which describe the different heat fluxes and represents temperature variations. This is implemented using the system dynamics paradigm, by modelling a stock which represents the temperature for each differential equation. The in and out flows describe the continuous variations of these temperatures. Figure 5.5 shows the refrigerator model in the SD notation, showing the different parts of the model.

5.2.2 Implementation of the Social Behaviour: Door Openings

As described in Section 4.6.2, human behaviour influences the operation of the refrigerator and thus its load curve. Therefore, we integrate so-called social behaviour that represents the non-technical influences on the device. Occurrences of door opening are irregular over a day, but surveys allow us to know some of their statistical properties. A simple agent is created based on the average number of door opening per day and the average duration of an opening. This agent acts on the refrigerator by modifying δ_d which triggers the heat fluxes (warm air entering the inner cell) when a door is open. This has a direct impact on the internal cell temperature T_i. Through a rise of T_i, the evaporator plate is reheated more quickly as well, this switches the thermostat and starts to run the compressor earlier. These dependencies can be clearly seen in the implementation diagram of the refrigerator (Figure 5.5.

In order to model door opening, a discrete state-chart was used which has two states, one for the open and one for the closed door. A transition for opening the door is added, which is based on an exponential distribution and is fired in average $\mu_{openingsPerDay}$ per day. For closing the door, a timeout transition is taken, which creates random durations according to a normal distribution with $\mu_{doDuration}$ and $\sigma_{doDuration}$.

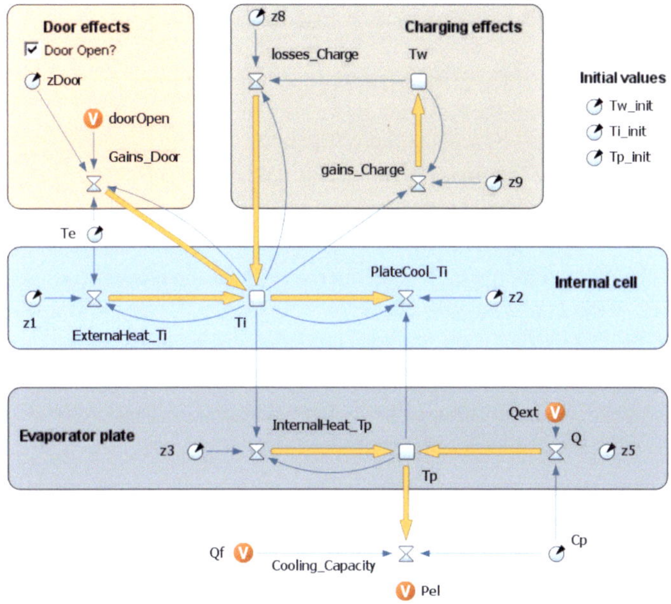

Figure 5.5: Refrigerator model implemented in system dynamics. The different parts of the system can be recognized, as well as the flows among them. Each stock (square) is defined by one of the differential equations given above.

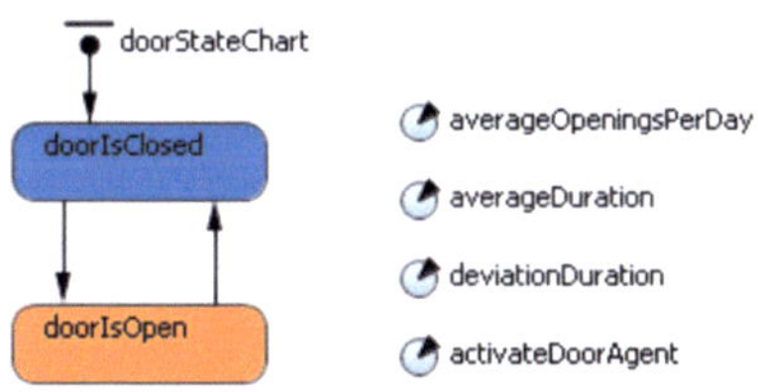

Figure 5.6: Door opening agent: This agents triggers door opening and adds a *human*, unpredictable behaviour onto the technical refrigerator model. On the left, the state-chart and on the right, the parameters used for the agent.

The values chosen by default were taken from a survey study [Laguerre et al., 2002]. An average of $\mu_{openingsPerDay}$ = 15 openings per day was taken, with an average duration of $\mu_{doDuration}$ = 20 s and a standard deviation of $\sigma_{doDuration}$ = 50 s. This creates many short door opening (such as for removing a bottle or some individual items) and fewer long door opening (such as refilling the refrigerator after shopping).

5.2.3 Demand Side Management Implementation through UFLS

As discussed in Section 4.7.2, under-frequency load shedding (UFLS) can help to improve the resilience of the system. A simple and distributed UFLS model is proposed here, which is implemented individually at the refrigerator level. The operation principle is as follows: the refrigerator is unplugged from the grid when a certain frequency threshold f_{off} is reached, and is reconnected when a second threshold f_{on} is passed. This aims to disconnect the load when there is a drop in frequency, usually caused by a production failure or a sudden, unforeseen load increase. The refrigerators are disconnected completely from the power source, so the compressor is no longer able to run. Once the frequency is stable, or some time has passed, they are reconnected again. Depending on the thermostat (acting on the evaporator temperature) the compressor may, or may not, begin to work again (see Equation 4.9). This algorithm is implemented as a state-chart which can be seen in Figure 5.7.

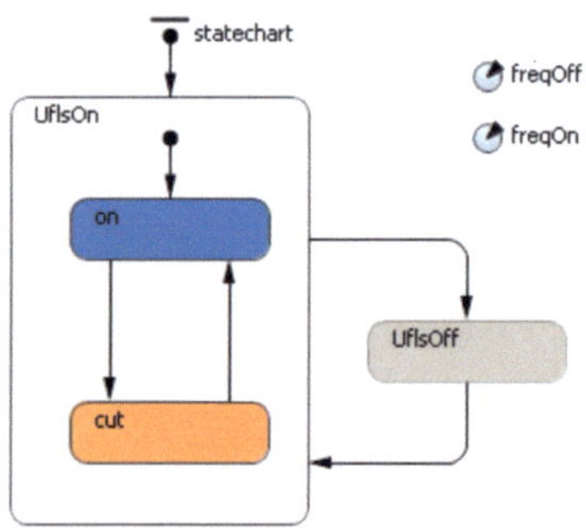

Figure 5.7: UFLS agent implementation: The UFLS model implements a device which can disconnect the refrigerator, based on the locally measured grid frequency. The *cut*-state disconnects the refrigerator from the grid. The state *UflsOff* serves to deactivate the UFLS agent and is triggered by the simulation GUI where the user can select whether UFLS is activated or not.

5.2.4 Measurement and Calibration of the Refrigerator Model

To analyse the effect of load shedding on a refrigerator, a real, typical domestic refrigerator was chosen and equipped with different sensors. The refrigerator was intentionally chosen to be an average and popular type, to be as representative as possible. Using a *new* refrigerator would not be sufficiently realistic, as the average installed refrigerator is an older model with lower performance, however because of difficulties in obtaining and equipping a used refrigerator, a new model was chosen. For future scenarios however, the selection of a new refrigerator was good, because in several years the refrigerator population will have been renewed and the sample refrigerator will better represent the refrigerators in common use.

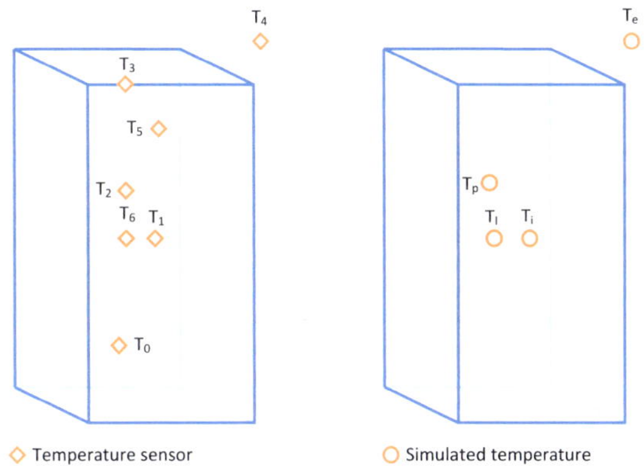

Figure 5.8: Temperature sensor locations and simulated temperatures in the model. On the left, the following measurements were taken: T_0 inside bottom, T_1 inside door, T_2 inside evaporator plate, T_3 inside top, T_4 room temperature, T_5 condenser plate, T_6 inside water middle. On the right, the simulated temperatures in the model: T_i inner cell, T_p evaporator plate (inside), T_l content (water), T_e room temperature.

The refrigerator was equipped with 7 temperature sensors in different places (Figure 5.8), both inside and out, in order to capture different temperatures. Furthermore, the active electrical power was measured, as well as the state of the door (open or closed) and the state of the plug (plugged or unplugged). Measurements were realised for different scenarios, such as empty operation without disturbances, or other more realistic scenarios including door opening and filling the refrigerator with food or drinks. By unplugging the refrigerator, the effect of disconnecting or performing a *load shedding* was analysed over different measured values. Once sufficient data was gathered (over several weeks), the model was calibrated using this data.

The calibration process was performed in different steps. First, the thermodynamics of the refrigerator cell and the plate without refrigeration were calibrated. Then, the parameters were re-calibrated against a regular cycle period at stable conditions. In a third step, the effect of door opening was calibrated by adjusting the door opening time constant τ_d. Furthermore, a long-term calibration was performed to adjust for behaviour over a period of a day and to adjust for long term trends (for example temperature recovery after food insertion, which takes several hours).

5.2.5 Integration of the SFR Grid Model

The model described in 4.8.2 relates to an unbalancing of production and demand to the corresponding frequency variation in a simplified way. The entries of the model describe a delta between actual production and the demand of the overall system, at a given point; it is called disturbance power. We define it generically as

$$\Delta P(t) = P_g(t) - P_d(t) \tag{5.1}$$

where P_g is the total generated power and P_d is the total demanded power. The original model is described in the block diagram formalism in the Laplace domain. This formalism can be directly implemented in, for example, Simulink; however, Anylogic, the modelling environment used here, includes the system dynamics paradigm. By transforming the equations back to the time domain, the equivalent differential equations can be found and implemented in SD as described in Appendix E.

The implementation of the SFR model in system dynamics can be seen in Figure 5.9.

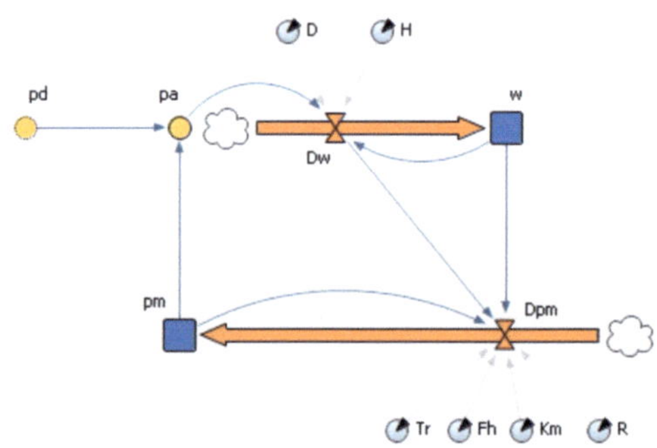

Figure 5.9: Representation of SFR model as a system dynamics stock and flow diagram

The two blocks of the original model (see Figure 4.7 on page 94) are translated into the two flows in SD which represent a first order differential equation. For the parameters, the following set was used (see Table 5.2. The two stocks ω and P_m represent the frequency variation and the mechanical turbine power, respectively.

Parameter	Description	Value
H	Inertial constant	4 s
D	Damping factor	1
K_m	Mechanical power gain factor	0.95
F_H	Fraction of total power generated by the HP turbine	0.05
T_R	Reheat time constant	8 s
R	Governor regulation	0.2

Table 5.2: SFR model parameters

5.2.6 Implementation of the Systemic Model

The configuration chosen in this case study is multi-approach, combining agent-based modelling with system dynamics. The refrigerators are represented as agents, and operate with an internal system dynamics model that simulates temperature dynamics and, from that, deduces power consumption. Furthermore, discrete events are used to manage the continuous equations, adding, for example door opening or load shedding events. These events modify the values of the continuous system dynamics equations at specific given moments.

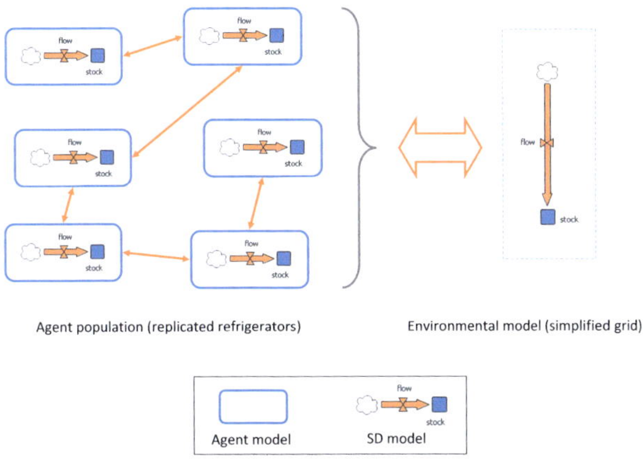

Figure 5.10: Multi-approach model chosen for the refrigerator case study: Refrigerators are modelled as agents, each one includes a system dynamic model (left). Interactions and operations are based on discrete event modelling. The energy system which calculates the grid frequency is also modelled in system dynamics (right) and coupled to the multi-agent model.

The agent-based model is coupled to an energy system grid model which simulates frequency response based on the balance of production and demand in the system. The simplified frequency response (SFR) model is implemented using the system dynamics paradigm as well. The approach can be seen in Figure 5.10.

Based on the population of refrigerators described, an integral model is created by (Figure 5.11) coupling the population to the SFR model. The energy system was represented through a simplified behaviour. The total production of the system was assumed to be constant, as well as the rest of the demand. In fact, the only variable loads in this test system are the refrigerators. This simplification allows us to analyse the direct impact of a refrigerator population on grid frequency, during an initial stage without taking into account the variations in power of all the other consumers. In a further stage, noise could be added to recreate the rest of consumer disturbances, but this was avoided intentionally in order to explicitly and clearly analyse just the effects of the refrigerator population.

The disturbance power is calculated here by

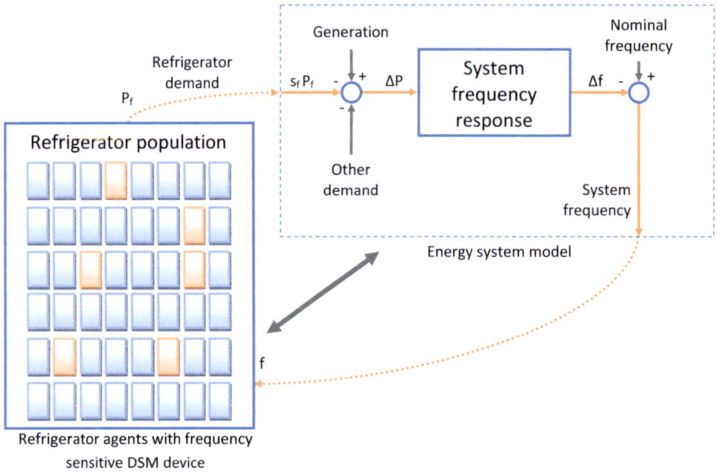

Figure 5.11: Integral system model: the multi-agent model of the refrigerator population interacts through power and frequency with the simplified energy system model. The aggregated load curve of the refrigerators affects the frequency of the system, which is again used at the refrigerator level in order to perform load shedding. The model represents the system interactions in a dynamic way, as both effects of frequency and power are inherently coupled.

$$\Delta P(t) = P_g(t) - P_d(t) - s_f \cdot P_f(f) \tag{5.2}$$

where P_g is the total generated power and P_d the consumption of all other loads and P_f the refrigerator load and s_f a scaling factor. A positive disturbance power therefore signifies over-generation, and a negative disturbance means an over-load of the system.

5.2.7 The Integrated Simulation Model

The graphical user interface (GUI) of the simulation model is shown in Figure 5.12. The simulation environment shows a graphic representation of each refrigerator, in which its state can be observed. The colour corresponds to internal cell temperature. Door opening is also visualised, as well as disconnection by the UFLS device.

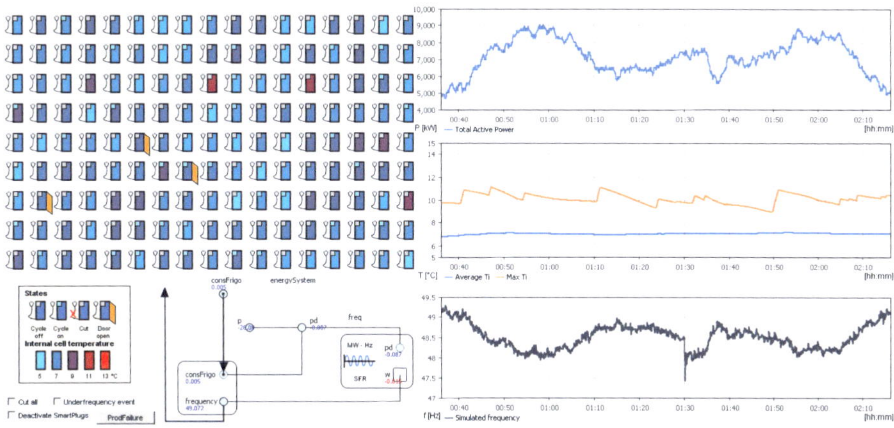

Figure 5.12: GUI of the refrigerator system simulation: on the left, the individual refrigerators are represented, showing their current state during simulation time. Below, the grid frequency model can be seen. On the right, plots allow the evaluation of the simulation results *in situ*.

Chapter 6

Simulation Results and Discussion

Having described the implementation of the models, they are ready for running simulations. In this chapter, the results for the two case studies on electrical energy systems are presented and discussed, taking into account their relevance to complexity. As we will see, many of the findings are related to characteristics and features of complex systems, which justify the chosen approach.

6.1 Multi-Timescale Simulation of a Wind Farm

In the following section, three simulations are reported in order to show the abilities of the model, to analyse the results and to assess the performance of the simulations. The first two studies are both simulated for a period of 24 h. The difference between them is that in the first case, a day with low wind speeds is simulated, whereas in the second case, high wind speeds are recreated. The third simulation is for a whole week, where (due to the duration) both high and lower speeds can be observed. The first two simulations allow us to analyse the reactions of the turbine park to low-speed effects such as the cut-in process when the wind is starting to blow. They also allow for the analysis of the effects on high speeds where cut-off phenomena can be observed. In the third week-long simulation, effects over a longer simulation period can be observed. In all cases, hourly and continuous simulations were run to compare the accuracy and performance of the models.

6.1.1 Low Wind Speed Day

In Figure 6.1 two plots are shown. In the upper plot, the wind speed as a comparison between hourly mean and continuous simulation is represented. The hourly mean wind speed, the interpolated hourly values and the simulated real-time speed (fast term) are shown in the first plot. The piecewise function of the un-interpolated hourly wind speed is the output of the slow term module. The interpolated hourly mean values are taken

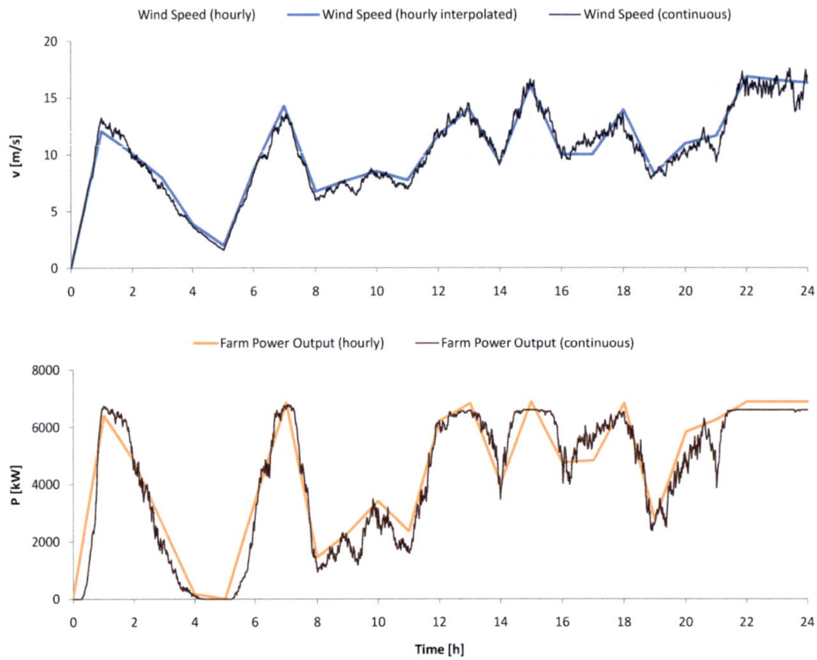

Figure 6.1: Comparison hourly and continuous power outputs (bottom) and corresponding wind speeds (top) for a day with low wind speeds.

from the linear interpolator. These are again used as input for the fast term module. The output curves of the wind farms are plotted below. Two curves are shown, one using the interpolated hourly mean speeds as inputs, and the second using the real-time, continuous wind speed output.

This first simulation shows a period of 24 h where wind speeds are relatively low, that means not exceeding 18 m/s. In particular, there are periods with low speeds, below 10 m/s, where a significant decrease of the output power of the turbines can be observed. Falling under the cut-in speed, they can even stop completely. The simulated wind farms are identical. The difference between them is the wind speed input data. The first farm takes the interpolated hourly mean wind speeds, the second one the real time speeds.

In Figure 6.1 we can see that the hourly computed power output of the farm follows more or less what could be a hourly mean of the continuous values. There are no great deviations, except a small one around 21h, due to a drop in the continuous wind speed caused by turbulence in the fast term.

Due to random failure behaviour, some differences caused by turbines in failure status can be observed (e.g. less total power in the last 2 hours of the day in the continuous simulation). It can be seen that the hourly power output approximately follows the continuous simulation, and only some short-term peaks are neglected (e.g. a drop in the wind speed at 21h that leads to a power drop is not visible in the hourly simulation).

6.1.2 High Wind Speed Day

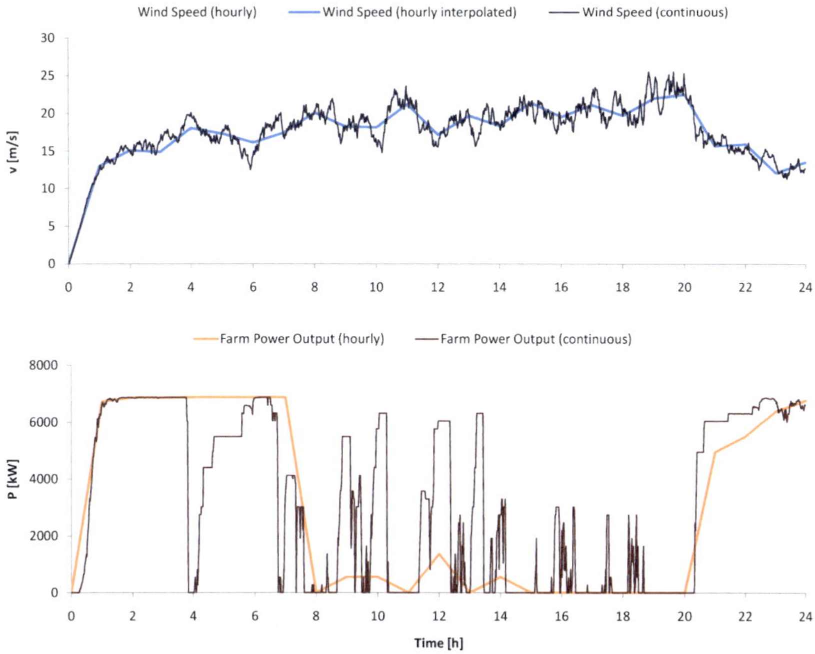

Figure 6.2: Comparison hourly and continuous power outputs (bottom) and corresponding wind speeds (top) for a day with high wind speeds. The cut-offs of the turbines can be clearly observed, especially for the continuous simulation, which makes the difference because of higher peaks in the latter.

In Figure 6.2 two plots can, again, be seen. Hourly and continuous wind speeds are represented on top while the aggregated electrical power outputs of the farm can be seen below. In this case, a day with high wind speeds was chosen. The speeds (once stabilised) are in the range of 12-25 m/s, being $v_{cut-off}$ = 20 m/s, so inside that range. Where the continuous wind speed is $v_w(t) > v_{cut-off}$, a cut-off for some or all (see Section 6.1.6) is achieved and they shut down, which leads to a complete power drop at the individual turbine scale, and important drops at the aggregated farm output. When $v_w(t) < v_{cut-back-in}$, the turbine starts again and causes a sudden power increase. These effects explain the strong fluctuations that can be observed in the continuous power output in the lower plot of Figure 6.2. It is interesting to observe the hourly output, too. There, no strong fluctuations are present, as can be clearly seen in the period between 8-20 h. Furthermore, cut-offs can occur in continuous output (due to the surpassing of the cut-off speed by some turbulences caused in the fast term module) which are not considered in hourly output, as the hourly mean remains $v_{h(t)} < v_{cut\text{-}off}$. This can be seen in the power drop between 4-5 h, while the hourly output stays at the nominal farm output. Thus, when dealing with fast speeds, the continuous model reflects strong fluctuations much better; these are neglected in the hourly simulation.

6.1.3 Simulation Over a Week

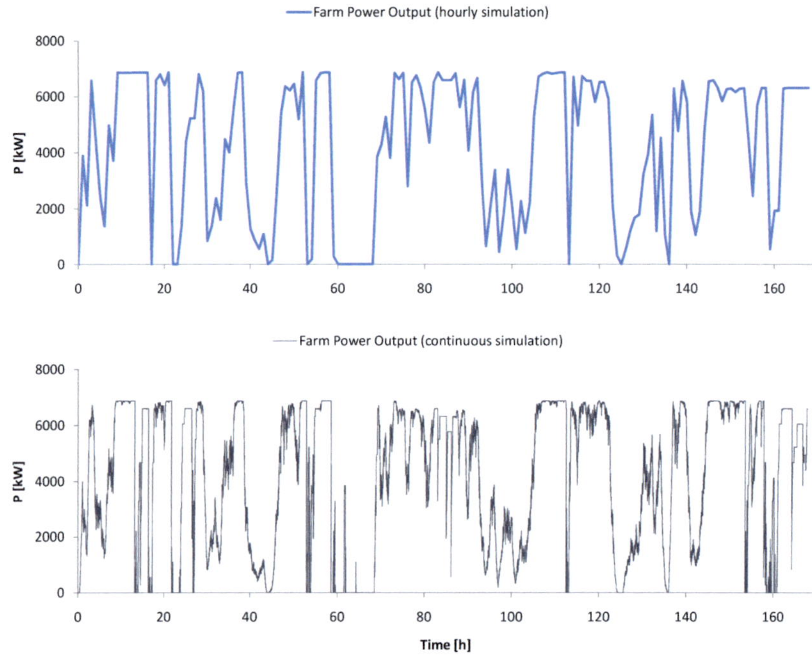

Figure 6.3: Comparison of a hourly (top) and a continuous (bottom) simulation for one week. It can be seen that the general trend is similar, but the continuous simulation shows some peaks and cut-offs that are not present in the hourly one.

In this case, a complete week was simulated. Figure 6.3 shows two plots of the power output for a 25 turbine wind farm (the same as in the previous examples), for hourly and continuous outputs, at top and bottom, respectively. As can be seen on the plots, over 7 days the output of each method differs strongly only in some cases. There are some points where $v_w(t) > v_{cut\text{-}off}$. The turbines shut down due to excessive wind speeds in this case, but looking at the same point in the hourly mean simulation, there is no such power drop. This is because $v_w(t)$ surpasses the hourly mean $v_h(t)$ punctually. To reach a power drop in the hourly simulation, $v_h(t) > v_{cut\text{-}off}$ is needed. These drops are a problem for grid stability, as they are very significant and occur in a short time frame. Indeed, control mechanisms for wind farms, to proactively shut down turbines based on wind speed forecasts or similar, in order to prevent such abrupt drops have not yet been considered in this model. Furthermore, the rapidly fluctuating wind speed component is transmitted to the power output of the plot below, while the curve of the hourly one is much smoother.

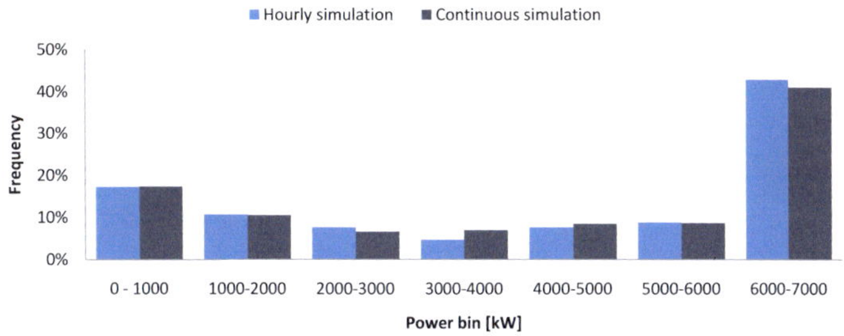

Figure 6.4: Histogram for the hourly and continuous simulations of the output power for one week.

In Figure 6.4, the histogram of both the continuous simulation (10 080 points) and the hourly average simulation (168 points) are compared. It can be seen that no large differences exist and the distribution is only slightly affected by the chosen method. For example, the high power values (6-7 MW) are more frequent in the hourly simulation, as some cut-offs are not considered in this model.

This is an example of how the model can be adapted to the different requirements of energy system simulations. If short-term data is needed, a real-time simulation can be run in order to get data that is continuous in time. If the simulation takes place over the medium term, i.e. some weeks or months, hourly mean speeds are used and the fast term component module is deactivated, giving a more efficient computation. For long-term simulations, the statistical data provided for the simulation can be used to compute the monthly energy output of wind farms.

6.1.4 Comparison of the Simulations

The simulations run above can be also compared on computational performance. In Table 6.1 a comparison of different features is shown. The use of only hourly mean value avoids the use of the fast-term component. This component is computationally slower, as it is based on a differential equation solver. By waiving this component, simulation performance can be significantly increased, (by around a factor of 50). However, it has to be taken into account that this increase is only affordable when accuracy and short-term fluctuation do not have to be considered (e.g. for longer term simulation). For simulating at higher temporary resolutions, though, the model including the fast term can be very interesting. Memory use is not considerably affected by the choice of the time resolution of the model.

Simulation period	24 h		168 h (1 week)	
Resolution	Continuous	Hourly	Continuous	Hourly
Number of turbines	25		25	
Execution time	122,0 s	2,3 s	753,8 s	14,5 s
Memory used	16 MB	16 MB	21 MB	15 MB

Table 6.1: Simulation run comparison

6.1.5 Failure Behaviour of the Turbine Units

As explained previously, the turbine model is provided with a failure function that allows us to simulate technical failures using specific parameters that can be obtained empirically. In this way, failures of individual units are randomly simulated over time. The average time to restart the turbines after such a failure is also considered.

Randomly driven timeouts are used to represent the transition to the failure state, which is triggered according to a rate. This rate is the inverse value of the MTBF. In order to get back to the working state, the rate corresponding to the MTTR is used. To trigger the transitions, exponentially distributed random numbers are used. The distribution is parametrised by this rate.

In Figure 5.4 the representation of the turbines and their current state is shown. The model can easily show the state of each turbine and the aggregated current output and energy production. The state of an individual generator and its production values can also be observed. The inclusion of failure behaviour in real time allows us to consider its direct influence on the power output of the farm, within the same model.

6.1.6 Heterogeneous Parameterisation of the Agents

All turbine manufacturers provide technical specifications that document their characteristics in detail. The values shown in these documents are not normally specified for each unit individually, as they are obtained using average values for all units of the same type. Although the units are supposed to be identical in construction, small differences cannot be avoided.

To model this heterogeneity, the turbine parameters were varied slightly within the population of turbines. The distribution used was a Normal distribution with a mean $\mu_{[value]}$ corresponding to the indicated value and a small standard deviation of $\sigma_{[value]} = 0.1\mu_{[value]}$. Further studies could obtain exact values for the variation of parameters among different units. This leads to small variations in the behaviour of each unit, that can result in aggregated effects on the wind farm output, and which are not usually considered in classical models. One of the strengths of the model is that it relies on the heterogeneous modelling of the individual agents.

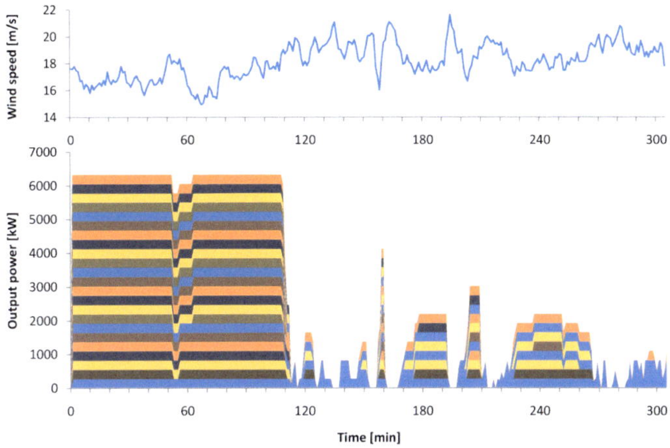

Figure 6.5: Simulated wind speed and total power output for a wind farm (continuous simulation) broken down by individual turbine. The wind speed is close to the cut-off threshold. Therefore, several stops and restarts of the turbines can be seen, which are not heterogeneous among the population.

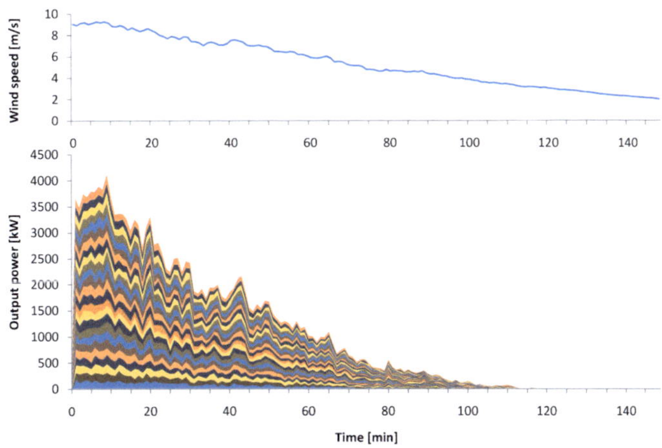

Figure 6.6: Simulated wind speed and total power output for a wind farm (continuous simulation) broken down by individual turbine. In this case, the wind speed is slowing down. The turbines reduce their output power in consequence.

Figure 6.5 shows the breakdown of the wind farm's power production by individual turbines. Their heterogeneous behaviour can be observed in this figure. Their different characteristics, as well as a slight variation of local wind speed, leads to unsynchronised turbine operation. This makes the model more realistic.

6.1.7 Discussion of the Results

Modelling the power output of wind farms at different time scales can be quite a complex activity. This case study presented a model for simulating a wind power system on multiple time scales. The multi-method approach was chosen in order to satisfy the various needs of the model, which integrates not only the pure generator but also failure behaviour and the consideration of a wind farm as a whole. Furthermore, the model was conceived to allow for simulation at different time scales, looking for the best computational efficiency in each case. The scalability included in this model allows the integration of different time scale simulations into the same module and the reduction of the number of total modules.

The model allows us to simulate wind power generation at different scales using the same model, by only switching between the different modules. The characteristics of the model are maintained at the different scales. So, for example, failure behaviour is modelled and can also be applied to short-term simulations, if needed. The following scope was made:

- The primary aim of the model is to simulate real time power outputs for energy system simulations rather than to estimate the accumulated energy productions over a period (used for example for the dimensioning of wind farms) .
- At high speeds, cut-off effects are better reflected in a high resolution (continuous simulation) model.
- The hourly model, though, can effectively approximate the hourly mean well in low speed periods.
- The model is sufficiently flexible to cover different needs arising from different time scales in integrated energy systems simulation.

This model brings together different modelling approaches, unifying continuous models, (differential equations, e.g. Equation (4.4), etc.) with discrete events (hourly changing mean speeds, state-chart modelling within the turbines) and agent-based modelling (e.g. of the failure behaviour and for the integration of the turbines into the wind farm). The use of different approaches allows for the creation of more realistic models that can take advantage of the different strengths of each approach. The agent-based approach, enables distributed parameters to be set to individual turbines, creating a heterogeneous park which recreates a more realistic behavior, both at the individual and at the aggregated scale. Furthermore, each turbine can be customised with real data (e.g. power curves, etc.). In this way, it is possible to simulate the realistic behaviour of wind farms in contrast to the static, homogeneous multiplication of identical objects.

A compromise between accuracy of the output powers and performance of the model must be determined depending on the application scope of the model. In large time-scale energy systems simulations, over several months or years, estimating wind speeds through the low term is sufficient. There is additional profit from the performance of this lightweight model. For the medium term, interpolated values can be used. For short-term simulations (up to some days) the fast term, which provides a model simulating high-resolution turbulences, can give better results.

Despite this, some drawbacks of the model were identified, among them wind direction, which is not taken into account for the moment. The model assumes that the turbines follow wind direction fairly well. The model is also only valid for active power injections, as reactive effects are not considered yet. In order to optimise continuous simulation it could be replaced by a minute-by-minute model, as the power output is not as directly coupled to wind speed as represented in the model, because of inertia of the rotor and modern automatic turbine regulation of the output.

Taking into consideration these limitations, and perhaps because of them), a simplified model that does not need large numbers of parameters was created. This allows for its integration to energy systems simulation as a light-weight, optimised model for different time scales.

6.2 Integrated Simulation of a Frequency-Based Refrigerator DSM

This case study goes one step further and performs simulations, connecting the multi-agent model of the demand side with the energy system model, which creates a dynamic, interacting systemic representation.

As in the example of the wind turbines, in order to represent heterogeneity among the refrigerators, each agent is parameterised in a different way. Furthermore, the human behaviour introduced by the door opening agent makes each refrigerator unique. By modelling each agent individually, no averaged or high-level model is used. This requires a greater computational power. It also allows for high-resolution models of the system, where individual values of each entity can be recovered. These values can be either individually or statistically analysed, over the whole, or part, of the population. In this case, the refrigerator agents will be replicated and will interact with an energy system model, which allows for the representation of grid frequency.

First, a simulation of the refrigerator population is discussed, which shows the effects on load curve and internal temperatures. Then the variation of the population size on the load curve is analysed. It makes quite a big difference, whether we consider only a few consumers or large populations of them. In order to quantify the effect of the size of the population, the resulting load curves are analysed.

In a second stage, under frequency load shedding (UFLS) is added to the refrigerators. This allows their power supply to be cut off when frequency drops are detected. The

UFLS strategy causes a rebound effect and a further synchronisation of the loads. This is analysed because complex system patterns were detected. Using the proposed UFLS strategy, oscillation emerges in the system. In a last step, the conditions for these emergent phenomena found in the simulations are determined. The phase transition, typical for complex systems, from a stable to an oscillating regime is discussed.

6.2.1 Simulation of a Refrigerator Population

In this simulation, normal operation of the refrigerator population is shown. 250 refrigerators are simulated with an installed power of 120 W each. All the parameters of the refrigerators are varied by 5% in order to recreate a realistic population in which each refrigerator has a slightly different configuration, as well as different initial conditions (internal temperature, etc.). This creates a heterogeneous population in which each refrigerator has its own state and therefore the individual load curves are particular to each agent. This is necessary for recreating the aggregated load curve, which is a composition of many individual peaks which, however, do not coincide. The average total load of the population is $\overline{P_f}$ = 11.5 kW during normal operation, which corresponds to a coincidence of around c = 40%.

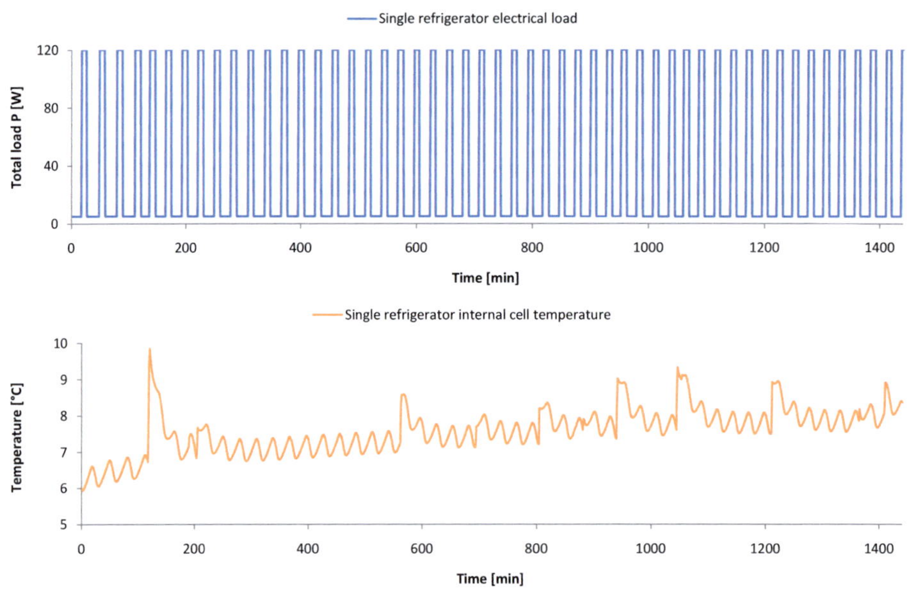

Figure 6.7: Simulation results for one randomly selected refrigerator of the population for one day. The load curve follows rectangular pulses due to the cyclic operation of the refrigerator. The temperature of the inner cell is maintained at around 7 - 9°C. The sharp increases in temperature are due to simulated door opening.

In Figure 6.7 the plots for one randomly selected refrigerator in the population are shown. The load curve is composed of a pulsing signal, which corresponds to the working periods of the compressor. In the temperature plot we see that the inner cell temperature is reduced during the compressor cycles and is increased when the compressor is not working. The sharp temperature increases are due to randomly created door opening by the door-opening agent, which acts on the refrigerator. After door opening, the working periods of the compressor are more frequent, because it must compensate for the increase in heat in the inner cell. In these circumstances the inner cell takes longer to cool down than when the door remains closed.

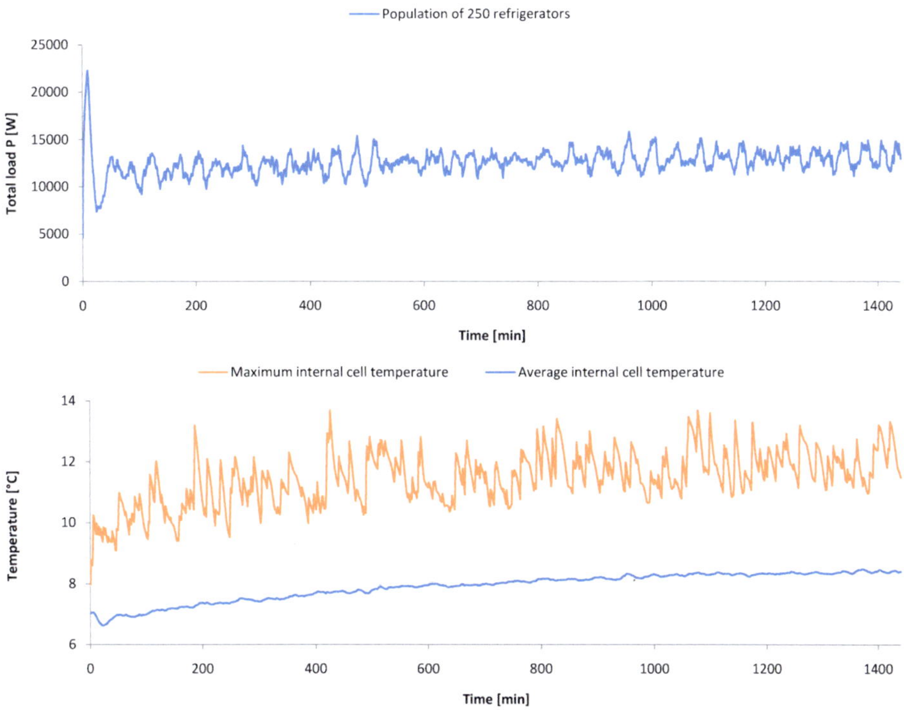

Figure 6.8: Simulation results for a population of 250 refrigerators for one day. The aggregated load curve is composed of many heterogeneous pulse-based curves which, when aggregated, do not coincide completely, this is why the curve flattens. The temperature plot below shows the average inner cell temperature of the whole population, and the maximum inner cell temperature of the warmest refrigerator at any given time.

In Figure 6.8, the total load curve for all refrigerators is plotted. We can see that, because the individual load coincidence is not perfect, the load curve has no sharp peaks. Each refrigerator has different parameters and door opening, and thus the pulses are not synchronised. This flat load curve represents approximately the average load, which can be estimated by the average operation coefficient of the whole population (which is around

40%) times the installed power. The temperatures shown are the average temperature of the inner cells of the whole population; this slightly increases over the day due to door opening. Door opening was not taken into account in the initial conditions. The maximum temperature is derived from the whole refrigerator population and represents the warmest refrigerator at any given time. This fluctuates significantly because door opening causes sudden increases in temperature.

6.2.2 Variation of the Population Size: the Aggregation Effect

Obtaining real world data for different population sizes in order to analyse aggregation effects of individual loads is quite a difficult task. In order to analyse many different scales of aggregation, the load curves of different sizes of populations have to be acquired. This requires significant effort because the analysis of many different scales necessitates individual load monitoring. This can be done with small-scale populations or, in the future, through the deployment of smart meters or similar technology, which allow data capture from individual consumers.

However, an agent-based simulation approach allows us to easily vary the size of the population and run many different experiments *in-silico*. In this case study, the effect of the variation of the population size of fluctuating consumers will be analysed on the aggregated curve. The refrigerators have a characteristic consumption based on periodic pulses, due to the compressor's intermittent mode of operation. An irregular consumption is a good example for exploring the aggregation effect, as continuous loads are easy to predict and do not present complex aggregation characteristics.

Having implemented the multi-agent simulation of a population of n refrigerators, different runs with different populations sizes were simulated. A larger number of refrigerators usually smooths the curve, as the coincidence of individual peaks is lower.

Figure 6.9 represents a simulation run for one day, for different sizes n of the population. It can be seen that the curve fluctuates less and is smoother as n gets larger. The cause of this is the dispersion of individual pulses that do not coincide and are better distributed in time. The effect that is observed between different number of agents will be called the *aggregation effect* in this work.

The population size was varied from $n = 5...50$ (in steps of 5) and $n = 50...250$ (in steps of 50) units. The load curve was analysed on the following features:

- Average load $\bar{P}$
- Standard deviation s
- Maximum load P_{max} per installed load P_{inst}
- Minimum load P_{min} per installed load P_{inst}

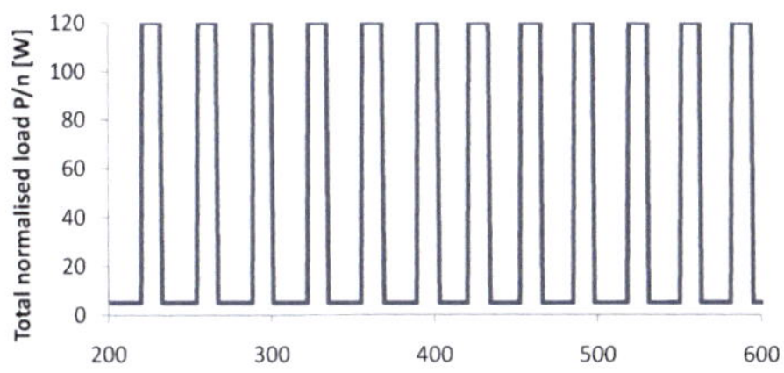

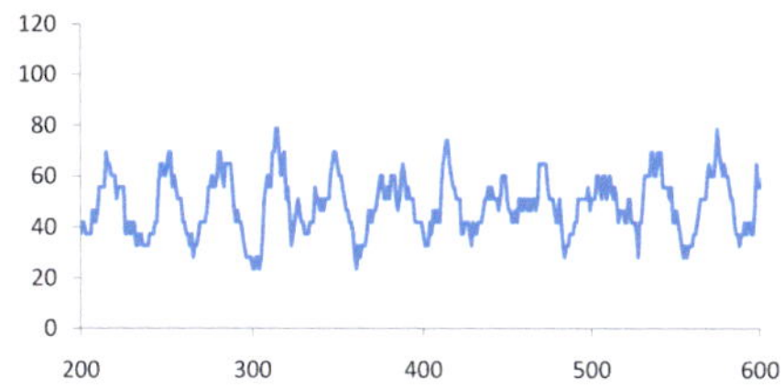

(a) $n = 1$

(b) $n = 25$

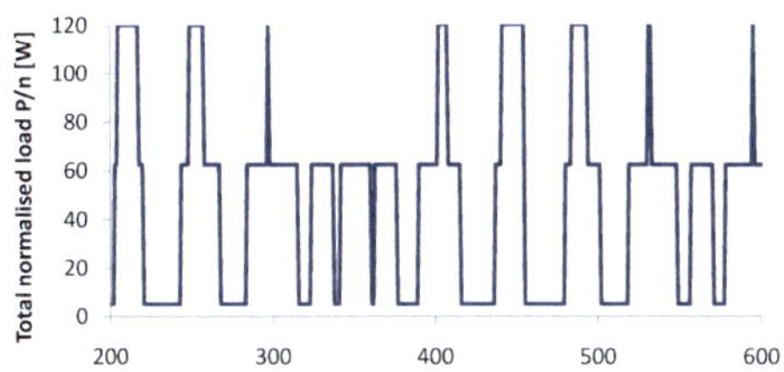

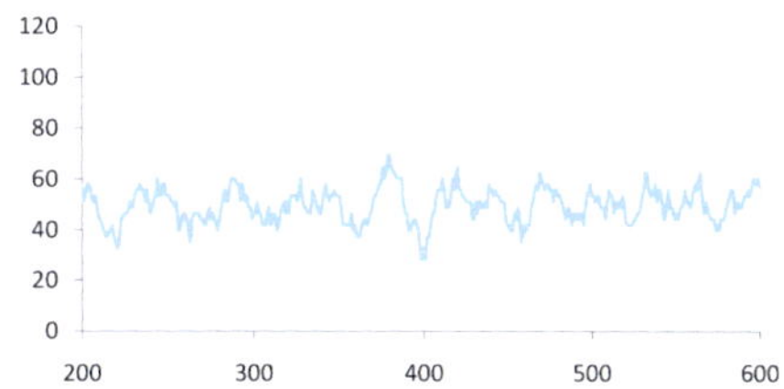

(c) $n = 2$

(d) $n = 50$

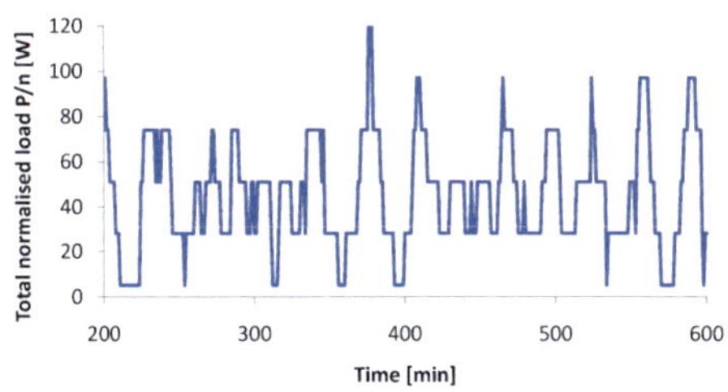

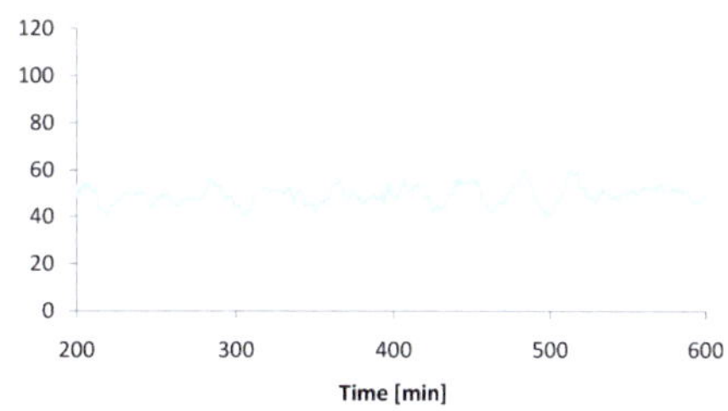

(e) $n = 5$

(f) $n = 250$

Figure 6.9: Aggregation effect for different population sizes: the load curve fluctuates depending on the population size n. For a small number of refrigerators, large fluctuations can be seen. The impact of one refrigerator agent is large, as can be seen by the rectangular pulses which are due to an individual agent for low n. For larger populations, the load curve flattens.

- Coincidence factor CF
- Load factor LF
- Execution time T

P_{max}, P_{min} and σ_P were normalised with n in order to be compared. The load curve gets smoother and varies less as the number of agents is increased. The average load is not affected by the population size. As it doesn't take into account fluctuations, even with small populations and the same refrigerator conditions, it is not dependant on the population size.

The standard deviation, or in this case, the coefficient of variation of the load curve decreases when the population grows. This is because, in small populations, the impact of one individual on the load curve is significant, and much sharper peaks appear than with larger populations. The standard deviation apparently follows a power law (see interpolated power-law trend line, which is quite well-matched to the simulation values). The standard deviation s thus seems clearly to be related to the population size. It can be approximated by:

$$s = a \cdot x^{(-b)} \qquad \text{with } a \approx 50 \text{ and } b \approx 0.5 \tag{6.1}$$

for the described case, with a coefficient of determination of $R^2 = 0.9819$. Similar findings appear when observing the evolution of the coincidence and load factors CF and LF, respectively. $CF = 1$ for very small populations, when the maximum load equals the total installed load. This means that there are moments where all single refrigerators are working at the same time. As we increase the number of refrigerators, the coincidence is lower, as it is much less likely that this occurs. The more refrigerators we have, the larger the difference becomes between the maximum reached power and the installed power. Again, we seem to be confronted with a power law function of n. The fluctuations are due to the stochastic nature of the model, which is important for small populations.

The load factor is a metric that relates the maximum load to the mean load. A low value indicates that the ratio of the maximum to the mean load is high, thus that it is far away from it. The value tends towards 1 as the curve gets flatter, which is the case for larger populations.

Indeed, CF and LF are closely related, as they both take into account the relationship of the maximum load to the total installed power and the mean, respectively. The mean power (which can be also obtained by the product of operation coefficient and installed individual power) is not dependant on the size of the population. The operation coefficient depends on the characteristics of the refrigerators, and again, not on its size. So, finally we can conclude that the variation of maximum power (or minimum power) to the mean follows a power law function of n.

The standard deviation is a measure of how widely values are dispersed from the average value (the mean). The standard deviation is also related but, as it is not calculated on one maximum but on the distances of all values from the mean, its statistical assertion is smoother. The maximum depends strongly on the non-determinism of the model, and varies more among simulation runs than the sum of the distances. As an average, though, and if repeating the same experiment many times, the coincidence and load factors should also remain stable and follow the power law.

The probability of extreme events, such as the coincidence of a large number of loads, should be further analysed. These events will strongly influence the load factor and co-

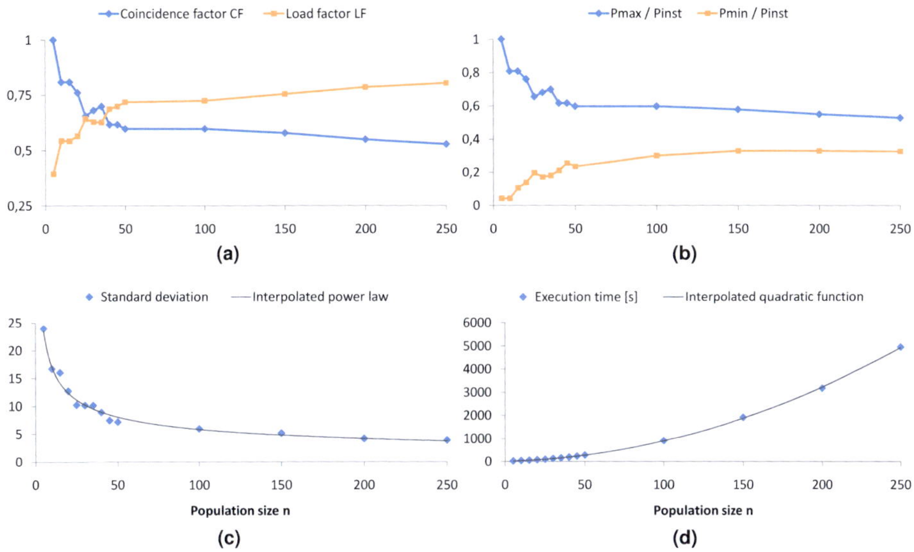

Figure 6.10: Aggregation effect and execution time: (a) shows that the coincidence and load factor are related to the size of the population through a power law, as well as the maximum and minimum power values (b). (c) shows the standard deviation of the simulations as a function of the population size. A clear power law dependence can be seen. The execution time which was plotted in (d) is related through a quadratic function to the population size.

incidence factor (as both are calculated considering the maximum peak load), but not the standard deviation.

Execution time depends on a polynomial relationship of the second order on the number of agents ($\Theta = n^2$, the measured times closely follow a quadratic dependency). This is due to the coupling of the model to the SFR frequency module, which creates an interaction among the agents. This increases the computational complexity which is no longer linear.

Stability of the simulation results Due to the non-determinism of the model, the smaller the population size, the higher the instabilities of the aggregated load curve results become. With large populations, the curve becomes more stable (being similar for different simulation runs), which is the aggregation effect. This is a simulation effect, but its parallel in reality can easily be drawn. If several measurements over a small population are taken, it is likely to that heterogeneous curves will be obtained. The influence of the behaviour of an individual is largely in relation to the total power. For the larger the populations, the fewer variations between different measurements will appear, if considered in similar conditions.

6.2.3 Simulation Settings for UFLS

For the under frequency simulation, the model is coupled with the frequency response system, which allows for the creation of an interaction between load and frequency. As in the first example, 250 refrigerator agents were simulated, parameterised heterogeneously with a variation of 5% of all parameters. In order to achieve an impact on a realistic energy system, the population load was scaled up by the factor $s_f = 1300$, which lead to an average scaled up load of 15 MW (for the refrigerators). The rest of the demanded load of the system was set to $P_d = 285$ MW, and the nominal generation power of the system to $P_g = 300$ MW. Therefore, the disturbance load was around some MW corresponding to the operation of the refrigerators.

6.2.4 An UFLS Scenario

Production failure is simulated through a sudden drop of 15 MW, which get restored linearly in the subsequent 10 minutes (see Figure 6.11). This recreates a failure in a plant with a subsequent reaction by the system (other generators take over the dropped load, following secondary reserve mechanisms). Production failure could be also generated by a sudden change in the wind speeds, as we have seen in the previous study. Both models were intentionally not coupled, however, as the complexity of dealing with both uncertainty on production and demand side management would overwhelm the possibility of analysing the results.

In the first simulation, no demand side management was performed on the refrigerator side. We can observe the aggregated load curve, which is rather smooth. Production failure causes the frequency to fall under 49 Hz for a short time, while the restoration of power stabilises it within the following 10 minutes.

Now we add an UFLS where a refrigerator is disconnected from the system completely, when the frequency falls under $f_{\text{off}} = 49$ Hz, and we reconnect it as soon as the frequency rises above $f_{\text{on}} = 50$ Hz. This is the first, most simple example of UFLS. As can be seen in Figure 6.11, as soon as the frequency falls below the threshold, the complete refrigerator load is released. However, after some time, the load of the refrigerators increases up to twice its average value (around 30 MW, which means that almost 80% of the refrigerators are working at the same time).

This is due to the principle upon which the refrigerators operate, the thermostat mechanism. A cut induces a kind of *reset* of the hysteresis process (see Equation 4.9) in some refrigerators, by interrupting their compressor cycle. This results in more refrigerators consuming power at the same time after the event, as the *natural* disorder of the system is perturbed by the disconnection.

The system frequency increases initially as a consequence of the UFLS disconnection, and within 6 s after the disconnection reaches 50.4 Hz. However, due to the increase in

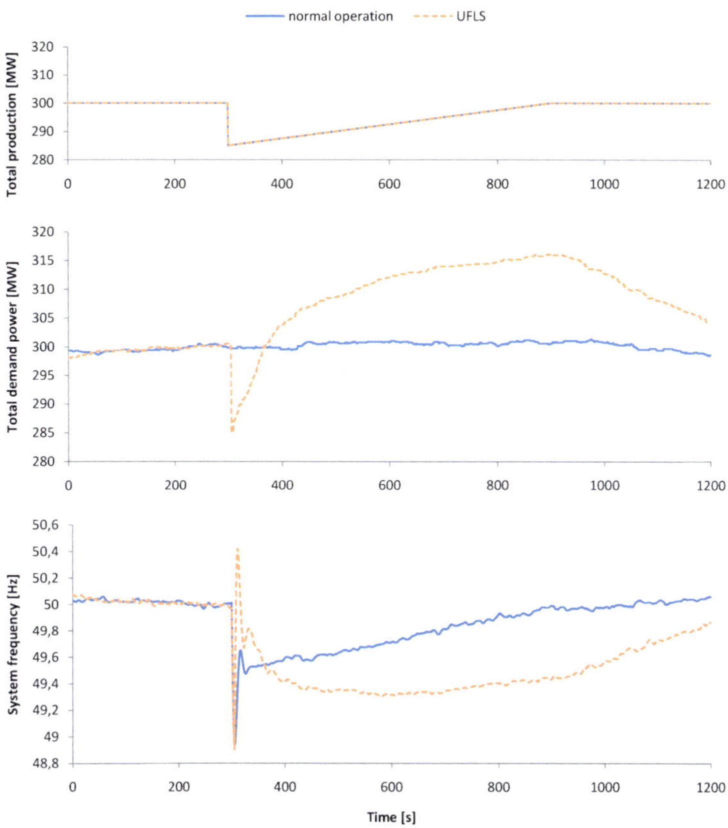

Figure 6.11: Simulation results for a production failure scenario with and without UFLS. In the simulation with UFLS, it can be seen how the demand drops down due to the shedding mechanisms of the refrigerators. Shortly after, frequency rises. Later, a rebound effect on the loads can be observed by the increase in power, in comparison to the normal case. The rebound lasts even after complete restoration of the generation side.

refrigerator consumption (in comparison to normal operation), the frequency recovery is slower and may even decrease again.

A very simple UFLS strategy based only on frequency thresholds can help to quickly restore frequency after an event, but it can also create a rebound effect due to the coincidence of loads after disconnection. This load increase is counter-productive to frequency stabilisation.

6.2.5 The Rebound Effect

As we have seen with UFLS, after the first reconnection, an increase in total load can be observed. This is usually called a rebound, which is characterised by a counteractive effect towards a measure taken before. This is worth detailed analysis, as the causes of a rebound are not always clear and can be complex.

An example of the rebound effect can be found in the introduction of more efficient devices for energy saving, which can induce an increase in consumption. There is an increase in use because of lower operational costs, resulting in maintaining, or even increasing, total consumption.

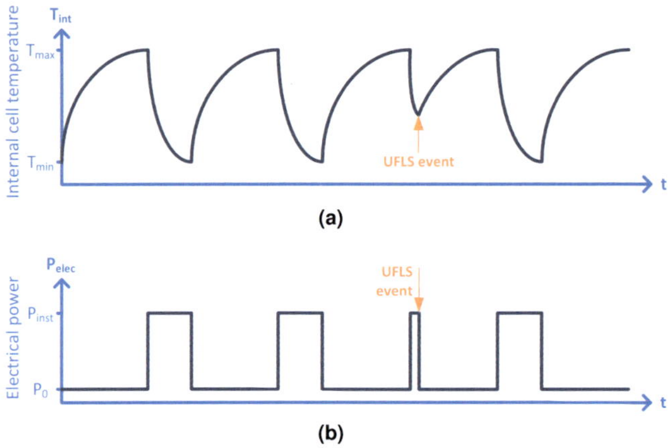

Figure 6.12: Operation cycles of a refrigerator: The internal temperature is plotted (a), and the correspondent load curve below (b). When a UFLS event occurs, the refrigerator is disconnected from the power supply and the cooling cycle is interrupted. When reconnecting, the refrigerator doesn't start working immediately, but only when the threshold temperature is reached again.

A rebound effect can occur on electrical load curves as a result of the recovery of energy lost during the shedding period. So, for example, if the refrigerator is cut from the power supply long enough, it will warm itself up. The power cycle, after reconnection, will be much longer than its corresponding thermal energy which has to be recovered. For a population of refrigerators, this means that the loads will coincide after load shedding.

When dealing with UFLS, the disconnection from the power supply only lasts some few seconds. This time is not relevant enough to have an effect on the internal cell temperature. However, rebound effect is also present in this case, even for these few seconds of disconnection. How can this be explained?

To understand why a rebound effect occurs on the aggregated curve, the operation mechanism of the refrigerator must be taken into account. In Figure 6.12a we can see the temperature curve which increases over a period of time until it reaches the T_{max} threshold. Then the compressor starts to work and the temperature decreases. The compressor won't stop until T_{min} is reached. The correspondent load curve is composed of periodic pulses. With an UFLS event, though, the cycle gets interrupted (if the event takes place during the compressor working state). When reconnecting, the compressor does not start to work immediately, but not until T_{max} is reached. This is the case for refrigerators working on this simple thermostat principle, described in the model in Equation (4.9).

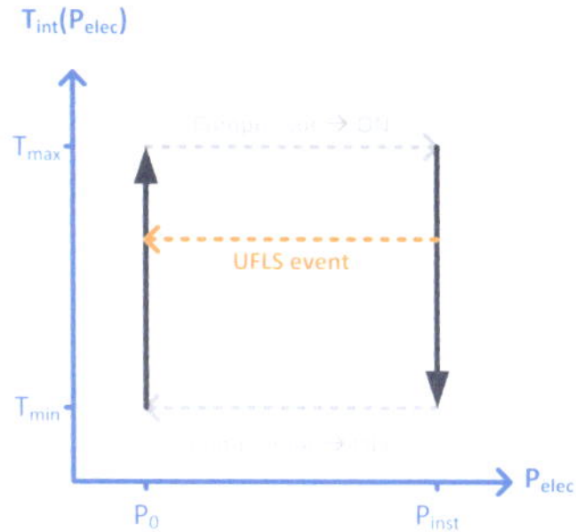

Figure 6.13: Hysteresis process for the refrigerator operation: the normal cycles are shown in dashed, grey lines. When reaching the thresholds, a discrete change in the system turns the compressor on or off. In the case of a UFLS event, a sudden disconnection occurs, causing an abnormal change of state (in orange).

This reset of the process can be better seen in Figure 6.13, where the hysteresis is shown in a phase plot of $T_{int}(P_{elec}$. A normal operation shows a discrete change when the compressor turns on or off, and the continuous temperature increases and decreases. The UFLS event causes the process to switch from the ON to the OFF state outside regular conditions. This reduces the time until the compressor restarts. For a population of refrigerators, this creates a larger probability of coincidence after an UFLS event.

The rebound effect is important as it is one of the main elements that causes a systemic phenomenon. After a UFLS event, a large increase in load induces a second frequency drop which, in this case is not caused by a production failure, but by an increase of load due to the rebound.

6.2.6 Synchronisation and Emergent Phenomena in the Simulations

Another scenario shows the effect of the same load shedding strategy when increasing the impact of refrigerator power on total demand. This was achieved by increasing s_f to 1760.

As the refrigerators are shed and reconnected after a short time (when frequency drops above 50 Hz), they tend to coincide in a higher degree, as previously described (Figure 6.14). The coincidence has shown to be greater after load shedding. Yet, no oscillation appears. As they reconnect again and again, however, a second under-frequency event takes place. This is not due to a production failure, but rather to a load increase by the refrigerators. The period gets closer, up to the point at which the system begins to periodically oscillate. This should obviously be avoided in a real system. The oscillation period seems to be related to the frequency response of the grid, and the characteristics of the refrigerator population. In the simulated cases, it is in the order of 2 - 4 seconds.

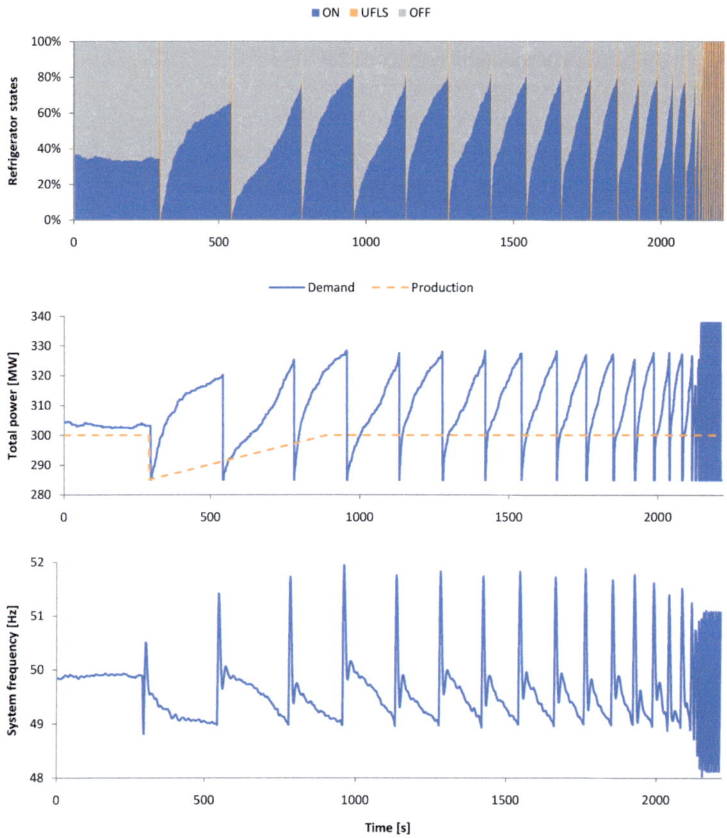

Figure 6.14: Simulation results for a production failure scenario with UFLS. The system begins to slowly synchronise itself up to a complete oscillation of frequency and loads.

Through the simulation model it was found that the main reason for the emergence of oscillation is the rebound effect which is created after the first disconnection. This rebound effect emerges due to the hysteresis process upon which the refrigerator thermostat works. The pulses tend to coincide more after each disconnection, as the thermostat cycles tend to synchronise. Similar effects have been described by Strogatz [2003] concerning fireflies and their light pulses, or synchronised hands clapping [Neda et al., 2000] in an audience.

The appearance of synchronisation apparently does not depend on the number of individual refrigerators, but rather on the aggregated load that is shed. We could reproduce the effect with a relatively small population of refrigerators (50-300) by scaling up the load. This means that, if the total load of the refrigerators managed by an UFLS is high enough in relation to the energy systems frequency response, a risk of synchronisation might exist when using a local response algorithm based on the same thresholds for every unit.

The time it takes for a complete system oscillation in this simulation is around 8 minutes after the first frequency event. However, it has been determined in further simulations that a complete oscillation of the system can take place sooner (after some minutes) or later (up to an hour), with the model configured in the same way. It depends on the state of the system at the moment of the under-frequency event, which is different for each simulation because of the stochastic models included in the agents.

Oscillation phenomena cannot be predicted from the individual models of the refrigerators. The phenomena emerges in a rather unstable way, which is related to the state of the population (describe by the individual states of each refrigerator), which has a very large degree of freedom. This is typical for complex systems. An agent-based model can recreate a population already able to represent the system as a whole, in a disaggregated way, with interactions among different levels.

6.2.7 Conditions for an Emergent Oscillation

Once having identified that the refrigerator system tends to oscillate in some cases, it is interesting to determine when this will occur. Several simulation runs showed that there is no discrete threshold between a stable and an oscillating system, but rather we are faced with a phase transition in which it is more or less probable that the system oscillates.

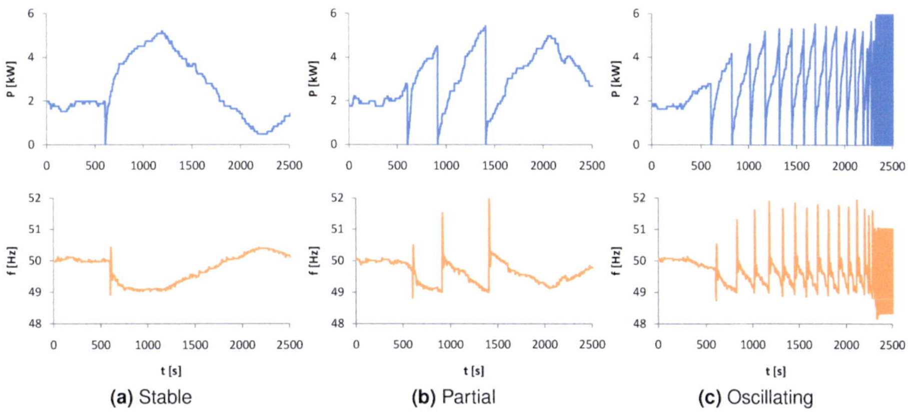

Figure 6.15: Examples of different regimes of the refrigerator system. On top, the total load curve of the refrigerators and below the corresponding grid frequency. The simulations were performed with the same parameters. Due to the non-deterministic nature of the model and for the given parameterisation, each of the three regimes can appear.

Based on the simulation scenario described so far, a Monte-Carlo evaluation was run to determine the probability of oscillation under different system conditions. First simulation runs suggested that the system oscillates when the total power is high enough to induce a frequency drop that underpasses the UFLS threshold frequency f_{off}. Therefore, for higher scaling factors, oscillation is more likely. The scaling factor s_f was varied from 6 000 to 12 750 in steps of 250. For each value of s_f, 100 simulation runs were executed. Three regimes were evaluated:

- Stable: the system responds to the under-frequency event by performing an UFLS only once.
- Partial oscillation: the system responds to the under-frequency event and begins to oscillate more than once, but does not achieve a complete oscillation. After some partial oscillations, the system stabilises again.
- Complete oscillation: the system begins to oscillate and the period becomes shorter up to the moment when it becomes constant.

The probability of each state as a function of s_f is shown in 6.16. As expected, it was found that the probability of staying in a stable system decreases for higher s_f values. However, the probability values of partial oscillation are still unstable. It increases up to a maximum of $p(\text{partial}) = 0.11$ and as the phase transition is passed, it decreases again for greater values of s_f. Even for 150 simulation runs per value, the partial state seems to be uncertain, and the results are not stable. Here lies the edge of chaos, between two phases whose outcome cannot be predicted.

The phase transition forms an s-shaped curve, which is common to other complex systems and appears, for example, in the diffusion processes (Bass model, diffusion of innovations, etc.). This s-shaped curve can be seen in Figure 6.16.

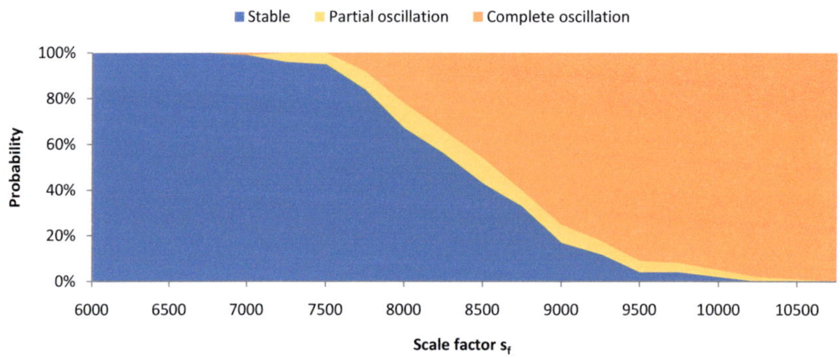

Figure 6.16: Conditions for a stable, partial and oscillating regime in function of the scale factor. The curve is obtained through a Monte-Carlo simulation, in which 100 runs were executed for each value of s_f.

6.2.8 Discussion of the Results

This example shows that there can be emergent phenomena and unstable states that are better understood when using complexity theory. Complex phase transitions take place

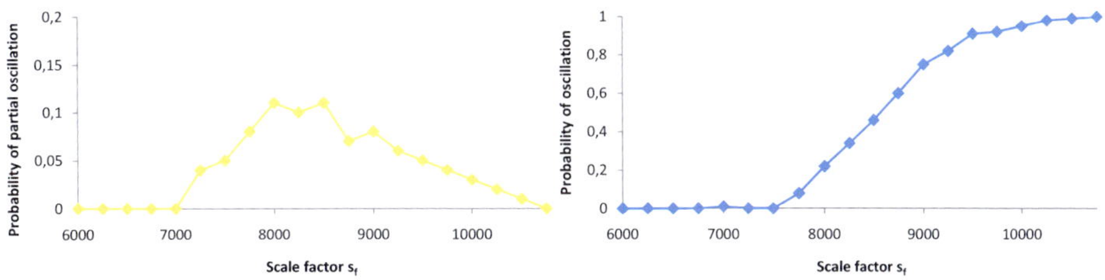

(a) Partial regime **(b)** Oscillation

Figure 6.17: Probability of a partial and a complete oscillation in detail: in (a), the partial oscillation regime can appear during the phase transition with a probability of around 10%. Its increase is faster on the stable side than its decrease on the oscillating side. In (b), the probability for a complete oscillation follows an s-shaped function.

even in this simple load shedding algorithm. The simulation model allows for their further exploration and for the quantification of the probability of a phase transition towards an oscillating state.

Moreover, a partial state was identified which overlaps with complexity itself. Behaviour is predictable both in the stable state and the oscillating state. However, in the intermediate state, it is uncertain whether the system will tend towards stability or oscillation. The Monte-Carlo analysis performed on the model confirmed this.

Knowing more about these phase transitions in real systems can help to further avoid unstable situations and to dimension a system that will stay in a stable state. The risk of a partial transition can be therefore reduced. As performing these kind of experiments on real systems is not possible, simulation can be a powerful tool in exploring behaviour in these conditions. For example, rare events can take place in a regime close to phase transition, but in the real system it would not be possible for sufficient experiments to be performed to attain this level of knowledge and the probability of an oscillation would be very low. However in this case, the risk exists and, through simulation, it can be identified and addressed.

Chapter 7

Conclusions and Outlook

The objective of this thesis was to show the interest and potential of a complex systems approach to the modelling and simulation of energy systems. This was first discussed from a theoretical point of view, presenting the fundamentals of complex systems, which are little known in the field of engineering. Complexity as an interdisciplinary science has been shown to help to understand complex processes in biological, sociological and other systems. Its application in the field of energy seems promising as similar patterns can be found, and as future energy systems will be more distributive in nature and include plenty of dynamic interactions, mainly because of the strong tendency to increase communication. Additionally, the energy system is multi-layered and is coupled to other non-technical aspects such as human behaviour, which requires a socio-technical approach.

The energy system has to be considered as a socio-technical system. However, this dimension is not yet part of many of the current simulation models. Including consumer behaviour in models is becoming more and more important as changes have to be correctly handled and can be handled at local levels. Behavioural changes can also be induced (by incentives) in order to drive the system into a particular trajectory (for example, incentive based DSM). Producer behaviour is also dependant on individual human decisions, such as the installation of domestic distributed generation, significant penetration of which could have a major impact on the whole system. Therefore, the inclusion of sociological aspects into systemic models is quite important.

The energy system is a true system-of-systems, made-up of a large number of subsystems that can act independently. This becomes even more relevant in the move towards a smarter, more distributed and decentralised grid that includes an intelligent control of subsystems and aggregators. The different subsystems are related to each other, by horizontal (within the same subsystem) and vertical (among different subsystems and scales) interactions.

Because of the need to represent a socio-technical, multi-scale system, a complex modelling approach was chosen. This is agent-based and allows for coupling with other modelling tools. Agents as autonomous, proactive and interacting entities can be described by continuous or discrete event models. The agent-based modelling approach allows a better understanding of a complex energy system. The chosen approach has an important explorative character. This means, that agent-based models can help explore many of the possible trajectories of a system before actually implementing it. Testing strategies beforehand and foreseeing potential future problems allows a more efficient conception and planning of future smart grids. The importance of the potential to include almost any kind of (inter and cross-scale) interactions should be underlined here. Agent-based models are naturally conceived in order to allow these kinds of interactions. Even if not already in the system, they can be easily added to the model and tested in simulation. This is especially important when considering an increasing degree of decentralisation and distribution of the energy system, which will encompass a larger number of interactions among its entities.

Using multi-approach modelling, in which agent-based models are coupled with continuous models, such as system dynamics or other differential equation modelling, seems promising. This approach is the only way to represent the complex electrical system in a disaggregated way. It allows for the representation of interactions between the individual entities of the system. These are important effects that cannot be captured by linear models so it is especially important to model the system with numerous interactions, according to the real system, in and between scales. Multi-scale modelling has been chosen as a fundamental approach, in which cross-scale relationships are represented. There are currently many models at different scales, which implement only horizontal interactions (among entities of the same scale). However, pertinent vertical interactions have to be identified and modelled too. This has been shown in the refrigerator case study, where individual loads can have effects at a system level (grid frequency) and create feedback loops. The indicators at a system level have a further effect at a lower scale of the system, which can create non-intuitive rebound effects, for instance.

The effects of topology and the relationship of complex network theory with electrical systems should be analysed further. As identified before, some transmission grids were shown to fulfil scale-free properties. An analysis of the distribution grid or other systems could reveal other interesting complex network properties. Furthermore, the analysis of networks in the communication layer and their relationship to the electrical layer should be considered. It seems obvious that topology plays a role in the design and operation of smart grids.

A number of different models of the electrical system were created. Following the philosophy of the method, these models are intended to be as simple as possible, in order to cover the needs of smart electrical systems simulation, but not more than that. During the modelling works, the importance of this step to simplify the models must be noted. There are many very good and detailed models on one hand, and, on the other, very simplistic models, which would be too inaccurate for our purpose. The process of taking a detailed model and reducing its degree of detail, in order to create a more simple one, is remarkably complex. It is important to know exactly the aim of the model in order to prepare it for that purpose. Furthermore, parameter reduction, in order to only take into account the really relevant values of the model, is not a trivial task. Therefore, real expertise in the area is needed. As we are dealing with many different fields, including electrical, mechanical, thermal, environmental and other disciplines, many discussions with different experts were needed in an atmosphere of interdisciplinary cooperation.

It should be underlined that the effort that is needed for the processes of simplifying models is commonly underestimated. This is especially so because of the perception that current tools, which implement detailed, non-systemic models are already sufficient and effective and need not be reduced. However, the need for a systemic approach in order to represent the system and its interactions requires exactly these type of models. They must be representative enough for the purpose, but also lightweight enough to allow large-scale simulations, in which many entities are executed in parallel and must allow for interactions.

Classical methods and models can also be combined with innovative approaches. The load profile method can complement agent-based models aiming to represent the demand side to complete or facilitate whole system modelling. Modelling each and every consumer individually might make sense in some cases, but might not be needed in others. For example, if only a specific electrical use is managed to perform DSM, it makes sense to represent this use in an agent-based form. For efficiency, the rest of the uses can be modelled by standard load profiles, as they will not be affected by the DSM measures. This still allows for the representation of total consumption. This is a good example of the importance of the level of detail necessary to describe a certain phenomena. Complex models of a *modular nature* allow for the different representation of the same reality, in different levels of detail. By using a modular approach as the one presented, depending on the question that has to be solved, an appropriate model for the considered entity can be chosen. Wherever possible, modular approaches were used. Modular properties are essential for creating large-scale systemic models which are easily extendable. In addition, *on-the-fly* modifications that allow dynamic change to the structure of the models during simulation time, must be possible.

There is still a significant need for standardisation in order to allow interactions between models. A framework is necessary to develop models to represent such a large and complex system as energy. A framework was developed (Tafat), in parallel to this work in cooperation with the University of Las Palmas, which aims to ensure compatibility between modules by defining the interfaces and structure of the models through a model driven engineering approach.

Another important point which was identified in the course of the work is validation. Validation of agent-based models of large-scale systems is a complex topic itself. It is too large a task to validate all the entities in these kinds of systems (and usually not all of the data needed is available). Calibration at one scale is not sufficient, as this only ensures validity at that scale, neglecting possible cross-scale effects. A calibration of the model is recommended at a minimum of two scales, in order to ensure validity of ABMs. Coherence between the two calibrations should be considered.

The two case studies presented in this work address these topics and illustrate, from a practical point of view, the aforementioned points. Based on these, the relevance of the approach is shown. Furthermore, some non-intuitive results were obtained which underline the interest of this modelling approach. Both the wind farm and the refrigerator case study represent multi-scale models. In the case of the wind farm study, multiple time scales were treated. A model with the possibility of switching the accuracy of its outputs was presented, which is modular and which can be used in different time-horizons for energy system modelling. The importance of heterogeneous behaviour and parameterisation was highlighted. If all the turbine agents are modelled identically, the output doesn't correspond to a real wind farm. Adding heterogeneous parameters and stochastic behaviour makes the simulation results more realistic.

In the second case study a multi-agent population of refrigerators is coupled with a simplified grid model, which allows the recreation of the behaviour of the production-demand balance on the frequency. The two scales are interacting dynamically with each other: the generated load from the refrigerators feeds the grid frequency model, and the frequency is then used to manage the refrigerators. A calibrated and individually validated refrigerator model was taken as a base. Using frequency-based demand side management allows the refrigerator to be *unplugged* at under frequency events marked by a threshold, and to be reconnected when the frequency surpasses a second threshold.

Some important and unexpected findings were made with this model. Firstly, the *aggregation effect* was analysed. This effect describes how a load curve varies according to the number of consumers. It explains why we obtain different load curves when measuring at an individual level, or at a low voltage transformer or substation level. As load behaviour over time is heterogeneous among individual consumers, the peaks do not coincide at the same time. Therefore, a flattening of the load curve occurs. However, up to this stage, very little is known about how loads *aggregate* and at which scale or number of consumers we obtain a representative, flattened load curve. Through the refrigerator example it was found that the standard deviation of the load curve follows a power-law, this means that there are high numbers of fluctuations for a small number of consumers, but for large numbers of consumers the deviation of the load curve is low. This relationship can be used to determine when the load curve is representative. Of course, we only studied the effect on refrigerators, but as power laws are common in complex systems it is likely that this type of relationship is valid for other cases also. This should be explored further.

The model allowed the testing of an under frequency load shedding strategy on a refrigerator population, and at the same time, the observation of the effects at a system scale. So, grid frequency was calculated in real time according to the consumption of the refrigerators. The frequency was used again to control load shedding at the individual scale. When using a simple load management strategy a rebound effect appears. Detailed analysis at the individual level indicates that this effect is due to the hysteresis process of the refrigerator thermostat and is not a result of the warming-up of the refrigerators during disconnection, as was first suspected. The model allowed us to explore these effects in detail. If required, a simulation can be stopped, and any of the values of the system (temperatures, DSM-state, thermostat state) observed and evaluated.

Furthermore, in some situations an emergent synchronisation appeared because the rebound effect caused another frequency drop. This synchronisation lead to a complete oscillation of loads and frequency, which would have been fatal in a real environment. This effect could only be observed due to the replication and simulation of a large and heterogeneous population of refrigerators, which for a *real* test bench would have been far too expensive. Emergent synchronisation has been studied in many other disciplines, and can be exemplified by firefly lightning or hands clapping. The effects found in the refrigerator population show many parallels to these studies and are worthy of further study. So, for example, we found certain values of parameters, which determine if the system remains stable if it synchronised partially or oscillates completely. This phase transition from one regime to another is typical for complex systems. In the intermediate stage, it is uncertain if the system will remain stable or if it will oscillate. This can be characterised by a probability, which follows an s-shaped curve and includes an unstable transitory state in-between. Knowing more about these phase transitions can help to avoid getting even close to a regime where a synchronisation would be possible (even with a very low probability), and occur as an extreme event. Systemic, cross-scale models like this can help to better understand emergent phenomena such as synchronisation or cascade effects in electrical energy systems.

A proposed simulation scenario would be to consider coupling both case studies presented in this work. The first and the second case studies could be coupled and disturbances to the system could be produced by wind generation. This would enable analysis of the reaction to a sudden drop in power caused by the wind speed. Furthermore, the strategies used for UFLS have to be improved, first to reduce the rebound effect and second to ensure a better frequency stabilisation and restoration. Therefore, different or slightly varied thresholds could be used.

Unlike in the biological or social fields, a much greater weight on controllability of emergence must be applied in energy systems. This is important because the other fields which use this type of approach are mostly interested *only* in greater analysis and understanding of the system's behaviour. However, in the case of a man-made energy system, which has been designed for a certain purpose, a deep understanding of its operation is needed; control and management of the system, including forecasting its behaviour are really fundamental. The complex systems approach allows an understanding of the operation of a potential future system. Experiences gained through simulation benefit the development of future systems. This approach offers an *in-silico* lab where upcoming technologies and strategies can be tested, in order to better manage the system, avoiding the costs and risks of real testing procedures. This is especially interesting in relation to distributed systems, in which large scale deployment would be necessary for testing. Strategies on how to manage these systems can be explored through a complexity-based approach. This approach enables the simulation of complex phenomena and interactions within the system, through different scales and with dynamic, heterogeneous entities evolving over time.

References

Advances in Complex Systems, Journal. 2010.
URL http://www.worldscinet.com/acs/mkt/aims_scope.shtml [cited on page 9]

ABRAS S. *Système domotique Multi-Agents pour la gestion de l'énergie dans l'habitat.* PhD thesis, 2009. [cited on page 56]

ADAMIC L. Complex systems: Unzipping Zipf's law. *Nature*, volume 474(7350):pages 164–165, 2011. 10.1038/474164a. [cited on page 25]

AKSOY H., FUAT TOPRAK Z., AYTEK A., and ERDEM NAL N. Stochastic generation of hourly mean wind speed data. *Renewable Energy*, volume 29(14):pages 2111–2131, 2004. [cited on pages 82 and 83]

ALBADI M.H. and EL-SAADANY E.F. Demand Response in Electricity Markets: An Overview. In *Power Engineering Society General Meeting, 2007. IEEE*, pages 1–5. 2007. [cited on page 35]

ALBERT R., JEONG H., and BARABÁSI A.L. Error and attack tolerance of complex networks. *Nature*, volume 406(6794):pages 378–382, 2000. 10.1038/35019019. [cited on page 26]

ALLERTON T. Simulating the distribution and cross-correlation of wind farm output. Technical report, EON and The Heriott-Watt University, 2008. [cited on page 82]

AMIN S.M. and WOLLENBERG B.F. Toward a smart grid: power delivery for the 21st century. *Power and Energy Magazine, IEEE*, volume 3(5):pages 34–41, 2005. 1540-7977. [cited on page 36]

ANDERSON P.M. and MIRHEYDAR M. A low-order system frequency response model. *Power Systems, IEEE Transactions on*, volume 5(3):pages 720–729, 1990. [cited on page 93]

APICELLA N. A Library for Decision-Making in Agent-Based Models. Fakultät für Informatik, Universität Karlsruhe (TH), 2009. [cited on page 58]

AUSTIN D. Percolation: Slipping through the Cracks. 2011. URL http://www.ams.org/samplings/feature-column/fcarc-percolation [cited on page 13]

AXELROD R. *The complexity of cooperation: agent-based models of competition and collaboration*. Princeton University Press, 1997. [cited on page 47]

BARABÁSI A.L. The Architecture of Complexity. *Control Systems Magazine, IEEE*, volume 27(4):pages 33–42, 2007. 0272-1708. [cited on pages 9 and 15]

BARABÁSI A.L. and ALBERT R. Emergence of Scaling in Random Networks. *Science*, volume 286(5439):pages 509–512, 1999. [cited on pages 15 and 21]

BAYEM H., PHULPIN Y., DESSANTE P., and BECT J. Probabilistic computation of wind farm power generation based on wind turbine dynamic modeling. In *10th International Conference on Probabilistic Methods Applied to Power Systems - PMAPS 2008*, pages 1–6. Porto Rico, 2008. [cited on pages 82 and 83]

BERGEN A.R. and VITTAL V. *Power System Analysis*. Prentice-Hall, New Jersey, 2000. [cited on page 92]

BILLINTON R., CHEN H., and GHAJAR R. Time-series models for reliability evaluation of power systems including wind energy. *Microelectronics and Reliability*, volume 36(9):pages 1253–1261, 1996. [cited on pages 82 and 83]

BOLLOBÁS B. and RIORDAN O. *Percolation*. Cambridge University Press, 2006. [cited on page 12]

BORSHCHEV A. and FILIPPOV A. From system dynamics and discrete event to practical agent based modeling: Reasons, techniques, tools. In *The 22nd International Conference of the System Dynamics Society*. The System Dynamics Society, Oxford, England, 2004. [cited on pages 66, 80, and 98]

BURIAN J. Complex Systems Tutorial. 2010. URL http://eldar.cz/cognition/complex/ [cited on page 49]

BURROUGH P.A. *Principles of geographical information systems for land resources assessment*. Monographs on soil and resource surveys ; no. 12. Clarendon Press, Oxford, England, 1986. Includes bibliographies and indexes.

CASTELLANI B. Art & Science Factory. 2012. URL http://www.art-sciencefactory.com/about-factory.html [cited on page 155]

CERRI G., PALMIERI A., MONTICELLI E., and PEZZOLI D. Identification of domestic refrigerator models including cool storage. In *International Congress of Refrigeration 2003*. Washington D.C., USA, 2003. [cited on pages 88 and 89]

CHAN W.K.V., YOUNG-JUN S., and MACAL C.M. Agent-based simulation tutorial - simulation of emergent behavior and differences between agent-based simulation and discrete-event simulation. In *Winter Simulation Conference (WSC), Proceedings of the 2010*, pages 135–150. 2010. [cited on page 73]

CHAPPIN E.J.L. and DIJKEMA G.P.J. An Agent Based Model of the System of Electricity Production Systems: Exploring the Impact of CO2 Emission-Trading. In *System of Systems Engineering, 2007. SoSE '07. IEEE International Conference on*, pages 1–5. 2007. [cited on page 73]

CHASSIN D.P. and POSSE C. Evaluating North American electric grid reliability using the Barabsi-Albert network model. *Physica A: Statistical Mechanics and its Applications*, volume 355(2-4):pages 667–677, 2005. 0378-4371 doi: DOI: 10.1016/j.physa.2005.02.051. [cited on page 26]

CHEDID R., AKIKI H., and RAHMAN S. A decision support technique for the design of hybrid solar-wind power systems. *Energy conversion, ieee transactions on*, volume 13(1):pages 76–83, 1998. [cited on page 85]

CHEN C.C., NAGL S., and CLACK C. A Method for Validating and Discovering Associations between Multi-level Emergent Behaviours in Agent-Based Simulations. In N. Nguyen, G. Jo, R. Howlett, and L. Jain, editors, *Agent and Multi-Agent Systems: Technologies and Applications*, volume 4953 of *Lecture Notes in Computer Science*, pages 1–10. Springer Berlin / Heidelberg, 2008. [cited on page 59]

COSTA GÓMEZ J. *Desarrollo de modelos de demanda eléctrica en el hogar*. Diploma thesis, 2011. [cited on page 85]

DEMAZEAU Y. and COSTA A. Populations and organizations in open multi-agent systems. In *1st National Symposium on Parallel and Distributed AI (PDAI96)*. Hyderabad, India, 1996. [cited on page 56]

DIMEAS A.L. and HATZIARGYRIOU N.D. A MAS architecture for Microgrids control. In *Intelligent Systems Application to Power Systems, 2005. Proceedings of the 13th International Conference on*, page 5. 2005. [cited on page 73]

DIMEAS A.L. and HATZIARGYRIOU N.D. Agent based control of Virtual Power Plants. In *Intelligent Systems Applications to Power Systems, 2007. ISAP 2007. International Conference on*, pages 1–6. 2007. [cited on page 73]

GONZÁLEZ DE DURANA J., BARAMBONES O., KREMERS E., and VIEJO P. Complete Agent based Simulation of Mini-Grids. In *The Ninth IASTED European Conference on Power and Engineering Systems, EuroPES 2009*, volume 681, pages 046–188. Acta Press, Palma de Mallorca, Spain, 2009. [cited on page 74]

GONZÁLEZ DE DURANA J.M., BARAMBONES O., KREMERS E., and VIEJO P. Modelling Smart Grids As Complex Systems Through The Implementation Of Intelligent Hubs. In *Proceedings of the 7th International Conference on Informatics in Control, Automation and Robotics ICINCO 2010*, volume 3, pages 146–151. INSTICC Press 2010, Funchal, 2010. [cited on page 74]

EHRENTREICH N. *Agent-Based Modeling : The Santa Fe Institute Artificial Stock Market Model Revisited.* Lecture Notes in Economics and Mathematical Systems ; 602. Springer-Verlag Berlin Heidelberg, Berlin, Heidelberg, 2008. In: Springer-Online.

EINSTEIN A. On the method of theoretical physics. *Philosophy of science*, volume 1(2):page 7, 1934. [cited on page 71]

EPSTEIN J. *Generative social science: Studies in agent-based computational modeling.* Princeton University Press, Princeton, USA, 2006. [cited on page 73]

EPSTEIN J.M. and AXTELL R. *Growing artificial societies: social science from the bottom up.* The Brookings Institution, Washington, DC, USA, 1996. ISBN 0-262-55025-3. [cited on page 73]

ERDŐS P. and RÉNYI A. On random graphs, I. *Publicationes Mathematicae (Debrecen)*, volume 6:pages 290–297, 1959. [cited on pages 14 and 19]

EUROPEAN COMMISSION. Microgrids Project. 2011.
URL `http://www.microgrids.eu` [cited on page 40]

EVORA J., KREMERS E., CUEVA S., HERNANDEZ M., HERNANDEZ J., and VIEJO P. Agent-based modelling of electrical load at household level. In S. Stepney, P.H. Welch, and P.S. Andrews, editors, *ECAL 2011: CoSMoS - Proceedings of the 2011 Workshop on Complex Systems Modelling and Simulation*, page 15. Paris, 2011. [cited on pages 74 and 86]

FERBER J. *Multi-agent systems: an introduction to distributed artificial intelligence*, volume 222. Addison-Wesley, 1999. [cited on page 51]

FERBER J. and PERROT J. *Les systèmes multi-agents: vers une intelligence collective.* InterEditions, 1995. [cited on pages 55 and 56]

FORD A. System Dynamics and the Electric Power Industry. *System Dynamics Review*, volume 13(1):pages 57–85, 1997. [cited on page 166]

FORRESTER J. *Industrial dynamics*. M.I.T. Press, 1961. [cited on page 62]

GILBERT N. *Agent-Based Models (Quantitative Applications in the Social Sciences)*. Sage Publications, Inc, 2007. [cited on page 53]

GIORDANO V., GANGALE F., FULLI G., and SÁNCHEZ JIMÉNEZ M. Smart Grid projects in Europe: lessons learned and current developments. Technical Report EUR 24856 EN, Joint Research Centre - Institute for Energy, 2011. [cited on pages 37 and 38]

GRIMMETT G. *Percolation*, volume 321. Springer Verlag, 1999. [cited on page 12]

GRÖNLUND B. Life and Complexity in Urban Space. 1993. URL `http://homepage.mac.com/bogronlund/get2net/Life_and_complexity.html` [cited on page 10]

HAREL D. Statecharts: a visual formalism for complex systems. *Science of Computer Programming*, volume 8(3):pages 231–274, 1987. [cited on page 65]

HATZIARGYRIOU N., DIMEAS A., and TSIKALAKIS A. Centralized and decentralized control of microgrids. *International Journal of Distributed Energy Resources*, volume 1(3):pages 197–212, 2005. [cited on page 73]

HAYEK F. The results of human action but not of human design. *Studies in philosophy, politics and economics*, pages 96–105, 1967. [cited on page 7]

JACOBO J.C., ROURE D.D., and GERDING E.H. An agent-based electrical power market. 2008. 14027501655-1656. [cited on page 73]

JENKINS N., ALLAN R., CROSSLEY P., KIRSCHEN D., and STRBAC G. *Embedded generation*. Institution of Electrical Engineers, 2000. (Power and Energy Ser. 31). [cited on pages 34 and 35]

JENNINGS N.R. On agent-based software engineering. *Artificial Intelligence*, volume 117(2):pages 277–296, 2000. [cited on page 60]

JIANG Z. Agent-Based Control Framework for Distributed Energy Resources Microgrids. In *Intelligent Agent Technology, 2006. IAT '06. IEEE/WIC/ACM International Conference on*. IEEE, 2006. [cited on page 74]

JIYUAN F. and BORLASE S. The evolution of distribution. *Power and Energy Magazine, IEEE*, volume 7(2):pages 63–68, 2009. 1540-7977. [cited on page 40]

JONES J.C., MYERSCOUGH M.R., GRAHAM S., and OLDROYD B.P. Honey Bee Nest Thermoregulation: Diversity Promotes Stability. *Science*, volume 305(5682):pages 402–404, 2004. [cited on page 13]

KOTTICK D., BLAU M., and EDELSTEIN D. Battery energy storage for frequency regulation in an island power system. *Energy Conversion, IEEE Transactions on*, volume 8(3):pages 455–459, 1993. [cited on page 93]

KREMERS E., LEWALD N., BARAMBONES O., and GONZÁLEZ DE DURANA J.M. An Agent-based Multi-Scale Wind Generation Model. In *The Ninth IASTED European Conference on Power and Engineering Systems, EuroPES 2009*, volume 681, pages 064–166. Acta Press, Palma de Mallorca, Spain, 2009. [cited on pages 74 and 82]

KREMERS E., VIEJO P., GONZÁLEZ DE DURANA J.M., and BARAMBONES O. A Complex Systems Modelling Approach for Decentralized Simulation of Electrical Microgrids. In *15th IEEE International Conference on Engineering of Complex Computer Systems*, page 8. Oxford, 2010. [cited on pages 67, 74, and 93]

KREMERS E., VIEJO P., LEWALD N., GONZÁLEZ DE DURANA J.M., and BARAMBONES O. Agent-Based Simulation of Wind Farm Generation at Multiple Time Scales. In G.O. Suvire, editor, *Wind Farm / Book 2*, volume 2. Intech, 2011. [cited on pages 74, 82, and 99]

LAGUERRE O., DERENS E., and PALAGOS B. Study of domestic refrigerator temperature and analysis of factors affecting temperature: a French survey. *International Journal of Refrigeration*, volume 25(5):pages 653–659, 2002. [cited on page 101]

LANGTON C.G. Computation at the edge of chaos: Phase transitions and emergent computation. *Physica D: Nonlinear Phenomena*, volume 42(1-3):pages 12–37, 1990. [cited on page 24]

LEE HAU AIK D. A general-order system frequency response model incorporating load shedding: analytic modeling and applications. *Power Systems, IEEE Transactions on*, volume 21(2):pages 709–717, 2006. [cited on page 93]

LIU Y.Y., SLOTINE J.J., and BARABÁSI A.L. Controllability of complex networks. *Nature*, volume 473(7346):pages 167–173, 2011. 10.1038/nature10011. [cited on page 28]

MACAL C.M. and NORTH M.J. Tutorial on agent-based modelling and simulation. *Journal of Simulation*, volume 4(3):pages 151–162, 2010. [cited on page 52]

MAIER M.W. Architecting principles for systems-of-systems. *Systems Engineering*, volume 1(4):pages 267–284, 1998. [cited on page 67]

MARKARIAN G., KÖLLE R., and TARTER A. *Aviation Security Engineering: A Holistic Approach*. Artech House Publishers, 2011. [cited on page 67]

MCARTHUR S.D.J., DAVIDSON E.M., CATTERSON V.M., DIMEAS A.L., HATZIARGYRIOU N.D., PONCI F., and FUNABASHI T. Multi-Agent Systems for Power Engineering Applications - Part I: Concepts, Approaches, and Technical Challenges. *Power*

Systems, IEEE Transactions on, volume 22(4):pages 1743–1752, 2007a. 0885-8950. [cited on page 73]

MCARTHUR S.D.J., DAVIDSON E.M., CATTERSON V.M., DIMEAS A.L., HATZIARGYRIOU N.D., PONCI F., and FUNABASHI T. Multi-Agent Systems for Power Engineering Applications - Part II: Technologies, Standards, and Tools for Building Multiagent Systems. *Power Systems, IEEE Transactions on*, volume 22(4):pages 1753–1759, 2007b. 0885-8950. [cited on page 73]

MILANO F. Power System Analysis Toolbox. 2009.
URL `http://www.power.uwaterloo.ca/~fmilano/psat.htm` [cited on pages 80 and 93]

MINSKY M. Matter, mind and models. *Proceedings of IFIP Congress*, pages 45–49, 1965. [cited on page 48]

MIORANDI D., ZANIN M., CRISTÓBAL S., HERRANZ R., and BALBÁS M. Complex Systems for an ICT-enabled Energy System (ComplexEnergy White Paper). Technical report, Innaxis Foundation, Create-Net, 2010. [cited on page 8]

NAVIGANT CONSULTING. Microgrids Research Assessment: Phase 2 Final Report. Technical report, 2006. [cited on pages 40 and 41]

NEDA Z., RAVASZ E., BRECHET Y., VICSEK T., and BARABASI A.L. Self-organizing processes: The sound of many hands clapping. *Nature*, volume 403(6772):pages 849–850, 2000. 10.1038/35002660. [cited on page 128]

NICHITA C., LUCA D., DAKYO B., and CEANGA E. Large Band Simulation of the Wind Speed for Real-Time Wind Turbine Simulators. *Power Engineering Review, IEEE*, volume 22(8):pages 63–63, 2002. 0272-1724. [cited on page 82]

OYARZABAL J., JIMENO J., RUELA J., ENGLER A., and HARDT C. Agent based Micro Grid Management System. *Power Engineering Society General Meeting, 2007. IEEE*, page 5, 2007. [cited on page 74]

PAATERO J.V. *Computational Studies on Variable Distributed Energy Systems*. PhD thesis, Helsinki University of Technology, 2009. [cited on page 33]

PACKARD N.H. and WOLFRAM S. Two-dimensional cellular automata. *Journal of Statistical Physics*, volume 38(5):pages 901–946, 1985. [cited on page 11]

PALENSKY P. and DIETRICH D. Demand Side Management: Demand Response, Intelligent Energy Systems, and Smart Loads. *Industrial Informatics, IEEE Transactions on*, volume 7(3):pages 381–388, 2011. [cited on page 35]

PALENSKY P., KUPZOG F., ZAIDI A.A., and KAI Z. Modeling domestic housing loads for demand response. In *Industrial Electronics, 2008. IECON 2008. 34th Annual Conference of IEEE*, pages 2742–2747. 2008. [cited on page 33]

PIPATTANASOMPORN M., FEROZE H., and RAHMAN S. Multi-agent systems in a distributed smart grid: Design and implementation. In *Power Systems Conference and Exposition, 2009. PSCE '09. IEEE/PES*, pages 1–8. 2009. [cited on page 74]

PYKA A. and GREBEL T. Agent-Based Modelling A Methodology for the Analysis of Qualitative Development Processes. pages 17–35. 2006. [cited on page 53]

RED ELÉCTRICA ESPAÑOLA. Wind power produced more than half the electricity in Spain during the early morning hours. 2009.
URL http://www.ree.es/ingles/sala_prensa/web/notas_detalle.aspx?id_nota=117 [cited on page 31]

REFSGAARD J.C. and HENRIKSEN H.J. Modelling guidelines–terminology and guiding principles. *Advances in Water Resources*, volume 27(1):pages 71–82, 2004. [cited on page 49]

REYNOLDS C.W. Flocks, herds and schools: A distributed behavioral model. *SIGGRAPH Comput. Graph.*, volume 21(4):pages 25–34, 1987. [cited on page 14]

RÍOS-RULL J. Models with heterogeneous agents. *Frontiers of business cycle research*, pages 98–125, 1995. [cited on page 72]

ROCHA L. Complex Systems Modeling: Using Metaphors From Nature in Simulation and Scientific Models. 2003.

ROSAS-CASALS M., VALVERDE S., and SOLÉ R.V. Topological vulnerability of the European power grid under errors and attacks. *International Journal of Bifurcations and Chaos*, volume 17(7):pages 2465–2475, 2007. [cited on pages 26 and 27]

SENSFUSS F. *Assessment of the impact of renewable electricity generation on the German electricity sector: an agent-based simulation approach.* Doctoral thesis, Universität Karlsruhe (TH), 2007. [cited on pages 72 and 73]

SHORT J.A., INFIELD D.G., and FRERIS L.L. Stabilization of Grid Frequency Through Dynamic Demand Control. *Power Systems, IEEE Transactions on*, volume 22(3):pages 1284–1293, 2007. [cited on pages 91 and 94]

SIMON H. *The Sciences of the Artificial - 3rd Edition.* The MIT Press, 1996. [cited on page 9]

SINGER J. Mental processes and brain architecture: Confronting the complex adaptive systems of human thought. volume 22, pages 1–1. Addison-Wesley Publishing Co., 1995. [cited on page 10]

SINGER J. Enabling Tomorrow's Electricity System - Report of the Ontario Smart Grid Forum. 2009. [cited on page 36]

SOLÉ R.V., ROSAS-CASALS M., COROMINAS-MURTRA B., and VALVERDE S. Robustness of the European power grids under intentional attack. *Physical Review E*, volume 77(2):page 026 102, 2008. PRE. [cited on page 26]

STROGATZ S. *Sync: The Emerging Science of Spontaneous Order*. Hyperion, 2003. [cited on page 128]

STROGATZ S.H., ABRAMS D.M., MCROBIE A., ECKHARDT B., and OTT E. Theoretical mechanics: Crowd synchrony on the Millennium Bridge. *Nature*, volume 438(7064):pages 43–44, 2005. 10.1038/43843a. [cited on page 14]

TARKOWSKI R. and ULIASZ-MISIAK B. Renewable energy sources in Guadeloupe. *Applied Energy*, volume 74:pages 221–228, 2003. [cited on page 42]

TELLIDOU A.C. and BAKIRTZIS A.G. Demand Response in Electricity Markets. In *Intelligent System Applications to Power Systems, 2009. ISAP '09. 15th International Conference on*, pages 1–6. 2009. [cited on page 73]

TERZIJA V.V. Adaptive underfrequency load shedding based on the magnitude of the disturbance estimation. *Power Systems, IEEE Transactions on*, volume 21(3):pages 1260–1266, 2006. [cited on page 93]

THIRIOT S. and KANT J. Generate country-scale networks of interaction from scattered statistics. In *Fifth Conference of the European Social Simulation Association (ESSA08)*. University of Brescia, Italy, 2008. [cited on page 61]

TOLBERT L.M., HAIRONG Q., and PENG F.Z. Scalable multi-agent system for real-time electric power management. In *Power Engineering Society Summer Meeting, 2001. IEEE*, volume 3, pages 1676–1679 vol.3. 2001. [cited on page 73]

TRANCHITA C., HADJSAID N., VIZITEU M., ROZEL B., and CAIRE R. ICT and Powers Systems: An Integrated Approach. In Z. Lukszo, G. Deconinck, and M.P.C. Weijnen, editors, *Securing Electricity Supply in the Cyber Age*, volume 15 of *Topics in Safety, Risk, Reliability and Quality*, pages 71–109. Springer Netherlands, 2010. [cited on page 74]

VALOV B. and HEIER S. Software for Analysis of Integration possibility of renewable energy units into electrical networks. In *The 5th International Conference Electric Power Quality and Supply Reliability*. Proceedings Tallin University of Technology pp 173-177., Viimsi, Estonia, 2006. [cited on page 80]

VAN DAM K. *Capturing socio-technical systems with agent-based modelling*. PhDthesis, Next Generation Infrastructures Foundation, 2009. [cited on page 79]

VANGHELUWE H. Foundations of modelling and simulation of complex systems. *Electronic Communications of the EASST*, volume 10(0), 2008. [cited on page 29]

VANGHELUWE H., DE LARA J., and MOSTERMAN P.J. An Introduction To Multi-Paradigm Modelling And Simulation. 2002. [cited on pages 48, 65, and 66]

VIEJO P. and KREMERS E. Simulation of energy system scenarios for regional planning decision-making using agent-based modeling. In *11th International Conference on Computers in Urban Planning and Urban Management CUPUM*. Hong Kong, 2009. [cited on pages 74 and 82]

VIEJO GARCIA P., GONŹALEZ DE DURANA J., BARAMBONES O., and KREMERS E. Quantitative system dynamics: Comparison of modeling techniques for the simulation of electro-mechanical systems. In *International System Dynamics Conference*. Washington, 2011. [cited on page 166]

VON NEUMANN J. and BURKS A.W. Theory of Self-Reproducing Automata. Technical report, University of Illinois Press, 1966. [cited on page 11]

WATTS D.J. and STROGATZ S.H. Collective dynamics of 'small-world' networks. *Nature*, volume 393(6684):pages 440–442, 1998. 0028-0836 10.1038/30918 10.1038/30918. [cited on pages 15 and 19]

WEISS G. *Multiagent systems: a modern approach to distributed artificial intelligence*. The MIT press, 1999. [cited on page 54]

WELFONDER E., NEIFER R., and SPANNER M. Development and experimental identification of dynamic models for wind turbines. *Control Engineering Practice*, volume 5(1):pages 63–73, 1997. [cited on page 83]

WIDEN J., LUNDH M., VASSILEVA I., DAHLQUIST E., ELLEGARD K., and WACKELGARD E. Constructing load profiles for household electricity and hot water from time-use data–Modelling approach and validation. *Energy and Buildings*, volume 41(7):pages 753–768, 2009. [cited on pages 33 and 91]

WIDEN J. and WACKELGARD E. A high-resolution stochastic model of domestic activity patterns and electricity demand. *Applied Energy*, volume 87(6):pages 1880–1892, 2010.

WILENSKY U. NetLogo Percolation model. 1998. Center for Connected Learning and Computer-Based Modeling, Northwestern University, Evanston, IL. [cited on page 12]

WILENSKY U. NetLogo. 1999. Center for Connected Learning and Computer-Based Modeling, Northwestern University, Evanston, IL. [cited on page 12]

WILENSKY U. NetLogo Life Turtle-Based model. 2005. Center for Connected Learning and Computer-Based Modeling, Northwestern University, Evanston, IL. [cited on page 11]

WILLIS H.L. and SCOTT W.G. *Distributed power generation: planning and evaluation.* Marcel Dekker, New York, 2000. [cited on page 33]

WOOLDRIDGE M.J. *An introduction to multiagent systems.* Wiley, Chichester, West Sussex, United Kingdom, 2009. [cited on pages 54, 55, and 60]

WRIGHT A. and FIRTH S. The nature of domestic electricity-loads and effects of time averaging on statistics and on-site generation calculations. *Applied Energy*, volume 84(4):pages 389–403, 2007. [cited on pages 33 and 91]

ZEIGLER B. *Multifacetted modelling and discrete event simulation.* Academic Press Professional, Inc. San Diego, CA, USA, 1984. [cited on page 48]

ZIMMERMAN R.D., MURILLO S., X, NCHEZ C.E., and THOMAS R.J. MATPOWER: Steady-State Operations, Planning, and Analysis Tools for Power Systems Research and Education. *Power Systems, IEEE Transactions on*, volume 26(1):pages 12–19, 2011. [cited on page 80]

Author's publications

APICELLA N., KREMERS E., and VIEJO P. A library of agent-based artificial intelligence in Anylogic modelling environment. In *23rd European Conference on Operational Research, EURO2009*. Bonn, Germany, 2009. Oral presentation. [cited on page 58]

BARAMBONES O., ALKORTA P., GONZÁLEZ DE DURANA J.M., and KREMERS E. A Robust Position Control for Induction Motors using a Load Torque Observer. In *IEEE 20th Mediterranean Conference on Control and Automation, MED'12*. Barcelona, Spain, 2012.

GONZÁLEZ DE DURANA J., BARAMBONES O., KREMERS E., and VIEJO P. Complete Agent based Simulation of Mini-Grids. In *The Ninth IASTED European Conference on Power and Engineering Systems, EuroPES 2009*, volume 681, pages 046–188. Acta Press, Palma de Mallorca, Spain, 2009. [cited on page 74]

GONZÁLEZ DE DURANA J.M., BARAMBONES O., KREMERS E., and VIEJO P. Modelling Smart Grids As Complex Systems Through The Implementation Of Intelligent Hubs. In *Proceedings of the 7th International Conference on Informatics in Control, Automation and Robotics ICINCO 2010*, volume 3, pages 146–151. INSTICC Press 2010, Funchal, Portugal, 2010a. [cited on page 74]

GONZÁLEZ DE DURANA J.M., BARAMBONES O., KREMERS E., and VIEJO P. Using The System Dynamics Paradigm In Teaching And Learning Technological University Subjets. In *International Conference on Education and New Learning Technologies EDULEARN 2010*, EDULEARN10 Proceedings, pages 2897–2906. IATED, Barcelona, Spain, 2010b.

EVORA J., KREMERS E., CUEVA S., HERNANDEZ M., HERNANDEZ J., and VIEJO P. Agent-based modelling of electrical load at household level. In S. Stepney, P.H. Welch, and P.S. Andrews, editors, *ECAL 2011: CoSMoS - Proceedings of the 2011 Workshop on Complex Systems Modelling and Simulation*, page 15. Paris, France, 2011. [cited on pages 74 and 86]

KREMERS E. *Untersuchung des Einsatzes von agentenbasierten Modellen für die Simulation dezentraler Energiesysteme am Beispiel von Guadeloupe*. Diploma thesis, Institut für Elektroenergiesysteme und Hochspannungstechnik, Universität Karlsruhe, 2008.

KREMERS E. An agent based approach for the modelling of integral energy systems. In *KIT PhD Symposium*. Karlsruhe Institute of Technology (KIT), Karlsruhe, Germany, 2009. Poster.

KREMERS E., GONZÁLEZ DE DURANA J.M., and BARAMBONES O. Emergent synchronization properties of a refrigerator demand side management system. *Applied Energy*, 2012a. Accepted on 18.07.2012. In press.

KREMERS E., GONZÁLEZ DE DURANA J.M., and BARAMBONES O. Synchronization phenomena in electrical systems: emergent oscillation in a refrigerator population. In *Complex System Design and Management, CSDM 2012*. Paris, France, 2012b. Accepted on 29.06.2012.

KREMERS E., LEWALD N., BARAMBONES O., and GONZÁLEZ DE DURANA J.M. An Agent-based Multi-Scale Wind Generation Model. In *The Ninth IASTED European Conference on Power and Engineering Systems, EuroPES 2009*, volume 681, pages 064–166. Acta Press, Palma de Mallorca, Spain, 2009. [cited on pages 74 and 82]

KREMERS E. and VIEJO P. Untersuchung des Einsatzes von agentenbasierten Modellen zur Simulation dezentraler Energiesysteme am Beispiel von Guadeloupe. In D. Möst, W. Fichtner, and A. Grunwald, editors, *Energiesystemanalyse*, pages 169–182. Universitätsverlag Karlsruhe, Karlsruhe, Germany, 2009.

KREMERS E. and VIEJO P. An agent based electricity demand model for the distribution grid. In *24th European Conference on Operational Research, EURO2010*. Lissabon, Portugal, 2010. Oral presentation.

KREMERS E., VIEJO P., GONZÁLEZ DE DURANA J.M., and BARAMBONES O. A Complex Systems Modelling Approach for Decentralized Simulation of Electrical Microgrids. In *15th IEEE International Conference on Engineering of Complex Computer Systems*, page 8. Oxford, UK, 2010. [cited on pages 67, 74, and 93]

KREMERS E., VIEJO P., LEWALD N., GONZÁLEZ DE DURANA J.M., and BARAMBONES O. Agent-Based Simulation of Wind Farm Generation at Multiple Time Scales. In G.O. Suvire, editor, *Wind Farm / Book 2*, volume 2. Intech, 2011. [cited on pages 74, 82, and 99]

VIEJO P. and KREMERS E. Simulation of energy system scenarios for regional planning decision-making using agent-based modeling. In *11th International Conference on Computers in Urban Planning and Urban Management CUPUM*. Hong Kong, 2009. [cited on pages 74 and 82]

VIEJO P., KREMERS E., TEJERA M.H., HERNANDEZ J., EVORA J., LANGLOIS P., DAUDE E., DURANA J.M.G.D., and BARAMBONES O. Towards an interdisciplinary approach for the simulation of future smart grid architectures from a complex systems science point of view. In *European Conference on Complex Systems (ECCS)*. Vienna, Austria, 2011.

VIEJO GARCIA P., GONZALEZ DE DURANA J., BARAMBONES O., and KREMERS E. Quantitative system dynamics: Comparison of modeling techniques for the simulation of electromechanical systems. In *International System Dynamics Conference*. Washington, USA, 2011. [cited on page 166]

Appendices

Appendix A

Complexity science map

Figure A.1 shows a *map* of complexity. It aims to represent a conceptual and historical evolution of complexity science and theory. The map is rough historical timeline divided into six major periods.

Each field of study is represented by a double-lined circle, with a double lined arrow moving from left to right. The double lined arrows represent the trajectory of each field of study. Areas of research identified for each field of study are represented with a bold, single-lined arrow. Leading scholars are included for each area of research. The font size demonstrates the relative importance of the authors within complexity science, according to Castellani [2012].

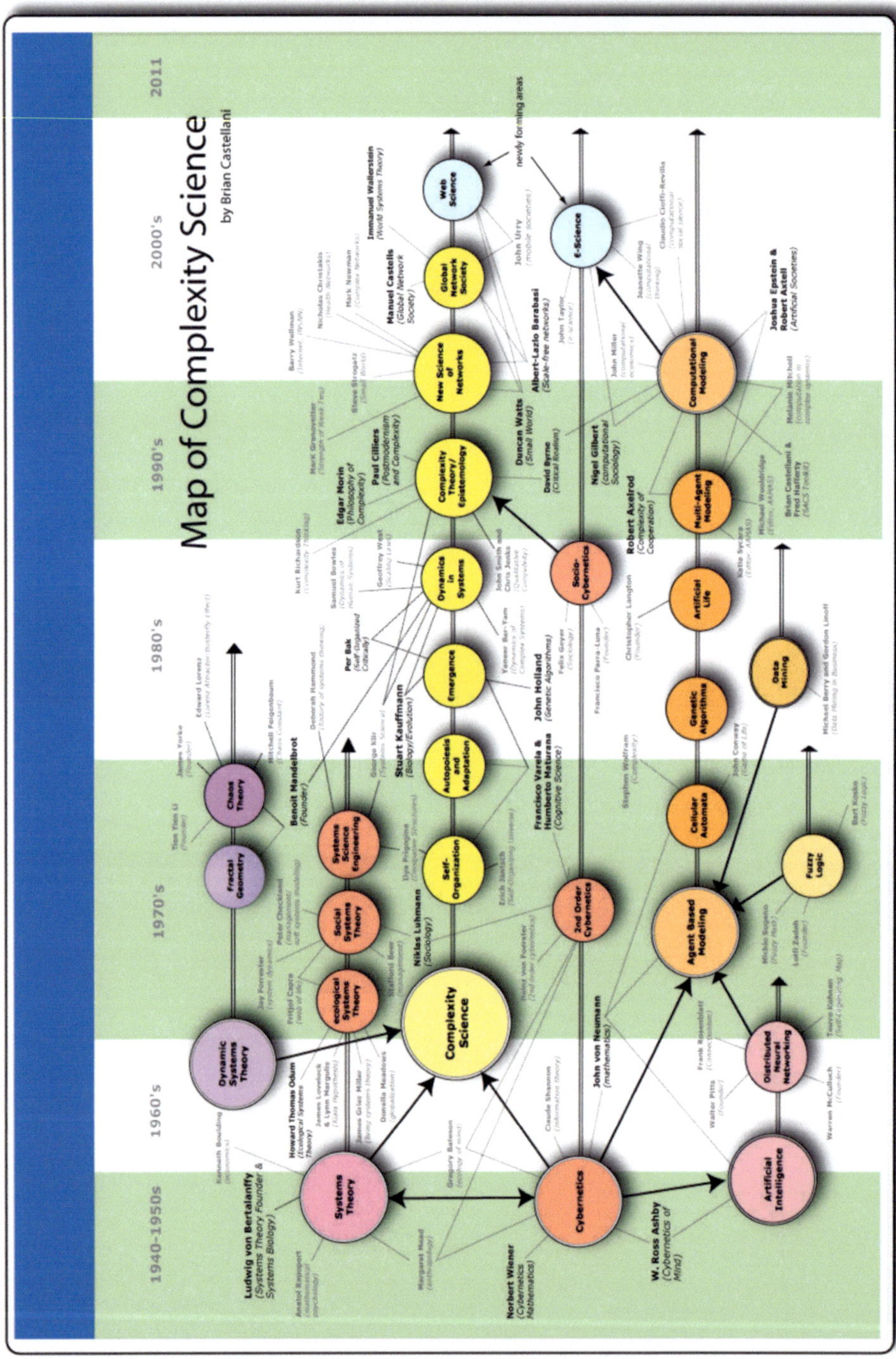

Figure A.1: Map of complexity science, related to social sciences. The figure represents a conceptual and historical overview of complexity science and theory. Source: `http://www.art-sciencefactory.com/complexity-map_feb09.html`

Appendix B

Technology Comparison for Storage Systems

Technology	Max Power [kW]	Ch./Disch. Time	E Density [Wh/kg]	Efficiency	Standby losses	Life time [cycles]
Supercapacitor	10	10^{-6}-1s	1-6 60?	90% 95%	0,1-0,2% / h	>1.000.000
SMES	7.000	1-20s	1,4	90% 95%	10-12% / day	1.000.000
Flywheel	15.000 50.000	10-30s 20-500s	4-60 12-140	85-90% 90-95%	3-20% / h 1-10% / h	1.000.000
Battery	17.000		20-160	70-90%	5-30% / month	100-2.000
Pb-Acid	17.000		20-50		5% / month	200-300
NiCd			25-80		5-20% / month	1000-2000
NiMH			55-120		30% / month	100-800
Li-Ion			90-160	70-90%	5-10% / month	300-1200
CAES	290 MW	2h	9	50-55% 70-75%		
Pumped Hydro	1.060MW	8,33h	8480MWh 1kWh/m^3	70-80%		
Fuel Cell	100MW		6.000	40-70%		
Hydrogen Storage		0,5h	10.000	30% 46%	0,3-1% / day	

Table B.1: Comparison of typical values for different storage technologies. Values in black are current, state-of-the art values; the values below in blue are expected values for technologies with a potential for improvement[1].

[1] Sources: *Umwelt Bundesamt: Zukunftsmarkt Elektrische Energiespeicherung (2007), TAB: Energiespeicher - Stand und Perspektiven (2008), BINE projektInfo (2003), FZ Jüllich*, `http://www.elektroniknet.de` and `http://www.cap-xx.com/resources/reviews/strge_cmprsn.htm`

Appendix C

Other Models

In this chapter, models relevant to energy systems which have been used in agent-based models are shown. These models are not included in this thesis but they could easily be integrated and combined with them through the modular approach.

C.1 Thermodynamic model of a household

The thermal behaviour of a building must be taken into account when estimating its electrical heating demand. Indoor air temperature (internal temperature θ_i) is linked to heat fluxes in the building. All heat fluxes must be taken into account, not only the flux through the envelope (transmissions and solar inputs) but also, internal thermal loads and air exchanges (due to convection or ventilation systems).

The energy balance is described by the following equation (based on Newton's law of cooling):

$$\rho c V \frac{\partial \theta_i}{dt} = \sum_j A_j \alpha_j (\theta_j - \theta_i) + n V \rho c (\theta_c - \theta_i) + \dot{Q}_{sol} + \dot{Q}_{ig} + \dot{Q}_h$$

where

- ρ density of air [kg/m^3],
- c specific heat capacity of air [J/(kgK)],
- V volume of the building [m^3],
- θ_i indoor air temperature [K] and
- t time [s];

the sum term corresponds to heat transmissions through the envelope of the building where for each surface j

- A_j surface area [m^2],
- α_j heat transfer coefficient [W/(m^{2}K)] and
- θ_j interior wall temperature of the surface [K];

the next term describes the air exchange due to air infiltration through leakage, the window and door frames or through intentional air exchange with the exterior (opening of windows or doors), where

- n air exchange rate [h^{-1}] and
- outdoor air temperature [K];

and the remaining terms represent other heat transfer rates due to

- Q_{sol} solar radiation input incident on the window surfaces which directly leads to an increase in temperature,
- Q_{ig} internal gains such as electrical appliances, human beings or lighting and
- Q_h heat contribution due to electrical heating.

This equation represents one thermal zone. Several zones could be represented in a building, but for simplicity's sake and, as the model will be reproduced in large scale simulation, the equation is considered once for the whole building (one zone model), thus disregarding different temperatures at different rooms and the heat transmissions between them.

The interior surface temperature was assumed at $\theta_j = \theta_i - 1$°C due to empirical data[1]. The solar radiation input was calculated according to the following equation.

$$\dot{Q}_{sol} = 0.8I \cdot 0.1A_w$$

where

- I solar irrandiance [W/m^2] and

[1]The exterior surface of the wall can also be taken into account by considering the heat transmissions: (interior air) - (interior surface); (interior surface) - (exterior surface); and (exterior surface) - (outdoor air). Therefore, the exterior surface temperature was assumed at $\theta_{j_e} = \theta_e + 2^{\circ}$ due to empirical data.

- A_w window surface area [m^2]

The correction factors 0.8 and 0.1 consider a partial coverage and/or reflection effects, and the fact that only a portion of the windows are exposed to the sunlight at any given time. It is provided by a radiation simulator or a historical dataset.

Outdoor temperature θ_e can be taken from a random simulator or from historic data. Internal gains can be calculated in several ways, in great detail, through using technical models of appliances that contribute heat to the interior temperature, or more broadly by taking static curves determined by measurements or other models. It is important to say that internal gains are strongly correlated to household size, the total number of persons living in the building, the number of persons currently present, weather conditions, etc.

Heat contributions will be explained in the next section as they are controlled to manage and keep reasonable levels for θ_i.

In this model, the differential equation which describes the heat flows is modelled using system dynamics, based on the approach presented in Appendix D.

Appendix D

Modelling Agents using SD

The increasing computational power, as well as the rapid development of software and simulation programs now allows for the creation of models of complex technical systems in areas such as architecture, engineering, economics and business, telecommunications, networks and the Internet. The development of these complex models is expensive and requires the cooperation of groups of people from different disciplines, often with different academic backgrounds. What is required is to employ a methodology that allows for the easy and fast exchange of models and ideas.

On the one hand, traditional education in engineering schools is based on the use of mathematical equations which will be referred to as Dynamic Systems (DS). These systems are precise and usually deterministic, purely quantitative, according to the needs of engineering sciences. The objective of these models is to be able to give a concrete answer to a problem, unrelated to planning or strategic purposes.

On the other hand, System Dynamics (SD), is a modeling techniques which was originally created to better understand complex industrial processes. SD models are based on stock and flows and are more visual and intuitive, and were introduced in Section 3.3.Qualitative approaches are mainly, but not exclusively found in SD. It is also possible to find quantitative approaches, as will be explained.

When multidisciplinary teams are working on models, communication problems arise that may hinder team integration and result in a drop in performance. This is mainly due to training differences between the team members. So, while people coming from *hard* sciences like physicists, mathematicians or engineers, are used to addressing the problems of dynamic systems in terms of differential equations (initial value problem (IVP)), people in economics, biology, architecture and philosophy, feel more comfortable when reasoning using Forrester diagrams, system thinking and stock & flow diagrams. These two groups of people have been working for decades in completely disjointed fields; SD people rarely exchange their views with DS people because they simply speak different languages.

Regarding methodology, DS people mainly use block-based tools like Matlab/Simulink and physical modeling tools that allow two-way connections, such as Dymola. SD people typically employ System Dynamics modeling software such as Stella, Vensim, etc. Clearly, an important step to improve the effectiveness of model building in multidisciplinary teams is to try to find a working methodology in which both DS and SD people feel comfortable. Methods used by the SD people are better positioned than DS methods to become the common methodology, some reasons for this are:

1. SD methods are easier to use for those familiar with DS methods than the reverse.

2. SD systems are depicted as SD diagrams (Forrester diagrams) representing *explicit* differential equations (with derivatives appearing only in the left hand side of the equations). They can easily be resolved by numerical calculation solvers. DS systems usually show *implicit* differential equations which result in a set of *differential algebraic equations*, which are not so easy to solve, particularly for multidimensional, discontinuous and discrete systems.

3. There are significant difficulties in adapting DS to allow more freedom in building models (blocks, bond-graphs and sophisticated components). For example algebraic loops arise very frequently when modeling systems in Matlab/Simulink. This is noticeable especially in hybrid systems modeling (discontinuous ordinary differential equations) where the SD method clearly offers many advantages.

4. A main advantage is the existence of high quality object-oriented software, that integrates SD as well as other modeling approaches, and allows for multi-approach modeling and team integration tasks.

Approach	+	–
DS	conciseness reproducibility preciseness	limitations in: space, comprehension, notation non scalable
SD	comprehension scalability multidimensionality	not *classical* lack of standards unusual representation

Table D.1: Advantages and disadvantages of the approaches

D.1 Numerical ODE Solution using an SD model

An initial value problem (IVP) is an ordinary differential equation (ODE) with a given initial condition (in form of a specified value) of the unknown function at a given point in the solution domain. In physics and engineering resolving these kind of problems is common, as the differential equation describes a system which evolves with time according to the specified initial conditions. Using SD we can obtain the numerical solution of the first order *control differential equation* (initial value problem),

$$\begin{cases} x'(t) & = f(t, x(t), u(t)) \\ x(0) & = x_0, \end{cases}$$

Anylogic hyperarrays[1] This allows for the modeling of these multidimensional differential equations, using a single Forrester diagram, as can be seen in Figure D.1.

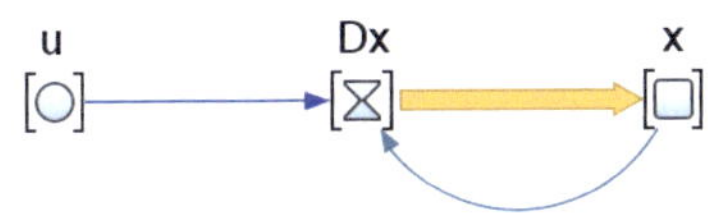

Figure D.1: Forrester diagram of an multidimensional ODE

Indeed, this diagram can represent a first order system when $n = 1$ or, in general, a system of order n system for $n > 1$ and then $u(t)$, $Dx(t)$ and $x(t)$ are hyperarrays. In this model, the (given) function $u(t)$ represents the system input (or control input). Dx as a flow variable in DS represents the formal derivative $x'(t)$ of the unknown function $x(t)$, which is represented as a stock.

The thick arrow is nothing more than the *integrator* DS object whereas the thin arrows indicate dependencies of the function

$$f(t, x(t), u(t)).$$

that is, f depends on x and u (also it depends implicitly on t).

D.1.1 First Order System

To explain a concrete example, a first order ODE given by

$$\text{IVP:} \quad \begin{cases} \dfrac{dx}{dt} = a\,x(t) + b\,u(t), \quad u(t) \text{ is given} \\ x(0) = x_0 \end{cases}$$

[1]A storage for multi-dimensional data used primarily in system dynamics models. Each data element is of Java type double. This construct (also known as *data with subscripts* or *array*) allows for easy and intuitive writing of formulae and equations, and flexible manipulating with dimensions. Arithmetic operations on arrays are performed element-by-element (unlike on matrixes in linear algebra).

where parameters a, b and control input are given. The following example is a special case from the previous example as its Forrester diagram is the same (but without array variable settings). The stock x variable properties are defined by

$$\frac{dx(t)}{dt} = Dx$$

as well as its initial value x_0. The Dx flow variable is described by the following function:

$$a\,x + b\,u$$

where a and b are Java variables of type *double*, declared within the Main window properties. Another double variable x_0 should be declared here.

D.2 SD Models in the Energy Sector

As shown previously, SD can help us to model basic engineering blocks for energy systems, for example RC or LC circuits which can be used to model power lines, or other basic elements of power systems, at a technical level. This application of SD is not the most common, and has been presented by the author in [Viejo Garcia et al., 2011]. It offers the advantage of using simple and schematic block representation to model technical systems in an environment like Anylogic. Also, the combination high and low level behavior by using the same formalism is advantageous. As seen in Section 3.3, SD is commonly used to represent systemic behavior at a higher level. At this level, a number of applications of SD in the energy sector can be found. [Ford, 1997] gives a review of existing models in the electric power industry. System dynamics has been used extensively to aid resource planning in this context.

Appendix E

Calibration of the Refrigerator Model

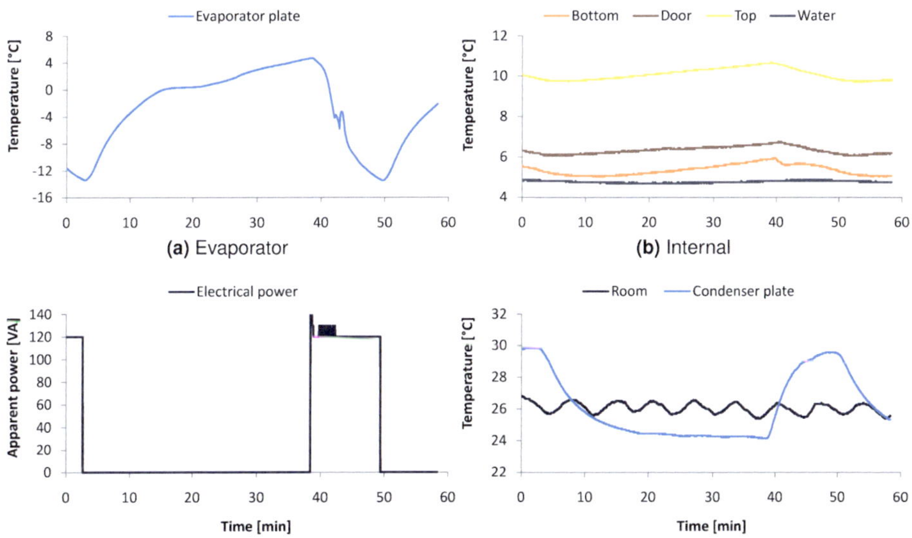

(a) Evaporator **(b)** Internal **(c)** Power **(d)** External

Figure E.1: Measured data for a stable cycle. The curves show the typical behaviour of the refrigerator in stable conditions, this means without door opening or changes of the contents. A periodic cycle can be seen, which consists on a warming-up phase, in which the compressor is off and the apparent power is almost zero, followed by a cooling-down phase in which the compressor is working and an almost constant apparent power can be observed.

The calibration of the refrigerator model is preformed in two major steps. First, the thermodynamics of the refrigerator cell and the plate, without refrigeration, are calibrated. Starting from a steady state after a refrigeration cycle, the evaporator plate is at its coolest temperature (around -13°C). Now, the refrigerator is unplugged and the dynamics of the temperatures are observed, which show, basically, the heat exchanges between the plate and the cell, as well as the cell and the exterior. In this first step, the cooling gains from the evaporator are intentionally not taken into account, in order to not calibrate too many parameters at a time. A charged scenario was contemplated for a more realistic behavior, as a refrigerator is rarely supposed to be completely empty. The test refrigerator was chosen to be charged with 12 x 1.5 l bottles of water, which would simulate an average filled fridge. The bottles were distributed throughout the cell and the door compartment. The temperature behavior observed in Figure E.1 shows the evolution of the different temperatures for one hour, which corresponds to one cycle, approximately.

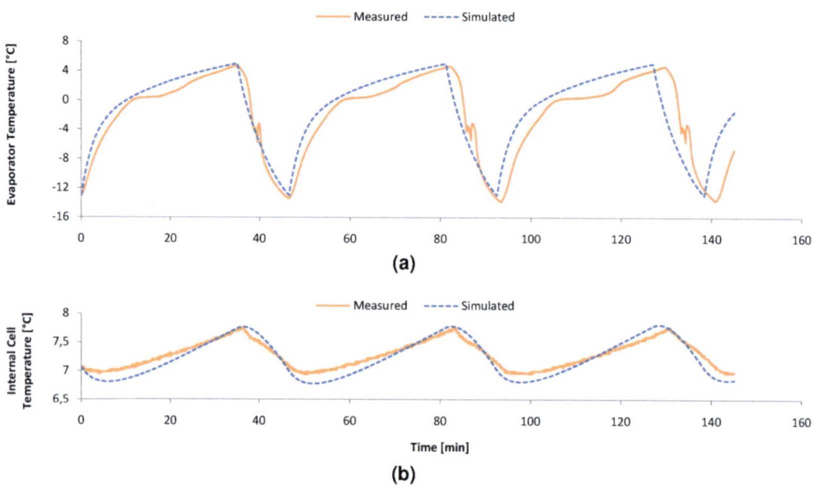

Figure E.2: Measured data and simulated values: (a) shows the measured and simulated temperatures of the evaporator plate. In (b), the inner cell temperatures seen. The simulated values for this calibrated model are a good approximation of the real data.

The calibration was performed using the OptQuest optimiser included in Anylogic. The optimiser was set to automatically vary the different time constants τ and Q_{ext} in a given range. By comparing the objective function (in this case the curve of the internal cell temperature and the evaporator plate temperature) with a target curves (the measured values), it tries to minimise the error by finding the most suitable parameter set. In this case, a double calibration was performed by fitting the evaporator temperature and the internal cell temperature at the same time. The error function E was defined as:

$$E = \sum_j (|T_i^{sim}(j) - T_i^{msr}(j)| + a \cdot |T_p^{sim}(j) - T_p^{msr}(j)|) \tag{E.1}$$

where T_i^{sim} is the simulated and T_i^{msr} the measured inner cell temperature and T_p^{sim} is the simulated and T_p^{msr} the measured temperature of the evaporator plate. The evaporator temperature varies around 16°C, the internal temperature around 2°C, therefore the evaporator error was weighted down by a factor $a = 0.2$.

In Figure E.2, the measured and simulated values can be seen, after calibration.

List of Symbols and Abbreviations

Abbreviation	Description	Definition
ABM	Agent-Based Model	page 52
ABS	Agent-Based Simulation	page 54
ABSM	Agent-Based Simulation Model	page 58
AC	Alternating Current	page 91
AI	Artificial Intelligence	page 56
AR	Auto-Regressive Model	page 82
ARMA	Auto-Regressive Moving Average Model	page 82
BA	Barabási-Albert Model	page 21
CAES	Compressed Air Energy Storage	page 44
CF	Coincidence Factor	page 34
CHP	Combined Heat and Power	page 31
CLD	Casual Loop Diagram	page 62
DSM	Distributed Generation	page 30
DSM	Demand Side Management	page 29
ER	Erdős-Rényi Model	page 19
EV	Electric Vehicle	page 37
GIS	Geographical Information System	page 78
GPRS	General Packet Radio Service	page 39
GSM	Global System for Mobile Communications	page 39
GUI	Graphical User Interface	page 107
HRES	Hybrid Renewable Energy System	page 40
ICT	Information and Communication Technologies	page 36
IT	Information Technologies	page 36
JRE	Java Runtime Environment	page 81
LF	Load Factor	page 34
MA	Moving Average Model	page 82
MAS	Multi-Agent System	page 51
MTBF	Mean Time Between Failures	page 85

Abbreviation	Description	Definition
MTTR	Mean Time To Recover	page 85
ODE	Ordinary Differential Equation	page 64
OOP	Object Oriented Programming	page 59
PCC	Point of Common Coupling	page 40
RES	Renewable Energy Source	page 31
SCADA	Supervisory Control and Data Adquisition	page 36
SD	System Dynamics	page 62
SFR	System Frequency Response	page 93
SMA	Système Multi-Agent (*French for MAS*)	page 54
SMES	Superconducting Magnetic Energy Storage	page 44
SOS	System of Systems	page 67
UCTE	Union for the Coordination of Transmission of Electricity	page 26
UFLS	Under Frequency Load Shedding	page 91
UMTS	Universal Mobile Telecommunications System	page 39
VPP	Virtual Power Plant	page 73
WMAN	Wireless Metropolitan Area Networks	page 39
WLAN	Wireless Local Area Networks	page 39
WS	Watts-Strogatz Model	page 20

List of Figures

List of Tables